Radio Frequency Identification: Technology and Applications

Radio Frequency Identification: Technology and Applications

Edited by
Ryder Wright

Larsen & Keller
www.larsen-keller.com

Radio Frequency Identification: Technology and Applications
Edited by Ryder Wright
ISBN: 978-1-63549-246-0 (Hardback)

© 2017 Larsen & Keller

▤ Larsen & Keller

Published by Larsen and Keller Education,
5 Penn Plaza,
19th Floor,
New York, NY 10001, USA

Cataloging-in-Publication Data

Radio frequency identification : technology and applications / edited by Ryder Wright.
    p. cm.
Includes bibliographical references and index.
ISBN 978-1-63549-246-0
1. Radio frequency identification systems. 2. Radio frequency identification
systems--Technological innovations. 3. Radio telemetry. I. Wright, Ryder.
TK6570.I34 R34 2017
621.384 192--dc23

The publisher's policy is to use permanent paper from mills that operate a sustainable forestry policy. Furthermore, the publisher ensures that the text paper and cover boards used have met acceptable environmental accreditation standards.

Printed and bound in the United States of America.

For more information regarding Larsen and Keller Education and its products, please visit the publisher's website www.larsen-keller.com

Table of Contents

Preface

Radio Frequency Identification (RFID) is used to track things and people by using electromagnetic fields. This technology is very advanced and some have local or battery power source. It is currently being used in industries like automobile industry, pharmaceutical industry, clothing, books, and is even being used in husbandry, as animals get implanted with the trackers. The book aims to shed light on some of the unexplored aspects of radio frequency identification. Most of the topics introduced in it cover new techniques and the applications of this technology. Different approaches, evaluations and methodologies and advanced studies have been included in it. As this field is emerging at a rapid pace, the contents of this textbook will help the readers understand the modern concepts and applications of the subject.

A detailed account of the significant topics covered in this book is provided below:

Chapter 1- Radio frequency identification works on the principle of electromagnetic fields that track tags or labels that have been attached to objects. The objects to be tracked have to be implanted with track tags and a reader is able to analyze the signals received from the tag. This chapter introduces the reader to the subject of radio frequency identification and radio frequency.

Chapter 2- Radio frequency is an electromagnetic frequency that lies between 3 kHz to 300 GHz. In the process of radio frequency identification, the tracking label exerts an electromagnetic field that is emitted in the form of radio waves that communicate data. The whole process use near-field communication to carry out the transmission of information. The chapter studies electromagnetic field, radio waves and near-field communication minutely.

Chapter 3- The main components of radio frequency identification discussed in the chapter are integrated circuit, modulation and demodulation. The integrated circuit forms the transmitter tag that relays modulated signals which are then demodulated to bring about the original information. The chapter explores the classification and advances in integrated circuits, modulation methods and demodulation techniques.

Chapter 4- This chapter deals exclusively with the tracking tags used in radio frequency identification. The various types discussed are clipped tag, bag tag, tochatag and ear tag. Each tag has a different design, composition, function and application. The reader is informed about these differentiating characteristics and the latest advances in each. The topics discussed in the chapter are of great importance to broaden the existing knowledge on radio frequency identification.

Chapter 5- The tracking abilities and ease of use makes radio frequency identification find application in numerous fields. Some of the applications discussed are biometric passport, electronic article surveillance, contactless payment, machine-readable passport, smartdust, transponder timing, telecommunication, intelligent transportation system and transponder. Tracking tags also find application in animal conservation efforts where it is implanted into target species to keep a track of their numbers and whereabouts.

Chapter 6- Radio navigation is used by flights and ships worldwide to receive position information transmitted from ground stations. Long-range radio navigation is not very accurate but short-range navigation can provide precise location within a few meters. The content covers topics like radiolocation, direction finding and fuzzy locating system. Radio navigation is best understood in confluence with the major topics listed in the following chapter.

Chapter 7 - Tracking systems offer the capability of providing real-time tracking and location management for various devices. The technology uses radio frequencies for computation and communication. Tools and techniques are an important component of any field of study. The following chapter elucidates the various tools and techniques that are related to radio frequency identification.

Chapter 8 - Data identification techniques refer to those devices that use radio frequencies to transmit and receive information as well as compute and make decisions with that data. The topics in this section discuss devices as well standards of data identification and processing. The aspects elucidated in this chapter are of vital importance, and provide a better understanding of radio frequency identification.

It gives me an immense pleasure to thank our entire team for their efforts. Finally in the end, I would like to thank my family and colleagues who have been a great source of inspiration and support.

Editor

Introduction to Radio Frequency Identification

Radio frequency identification works on the principle of electromagnetic fields that track tags or labels that have been attached to objects. The objects to be tracked have to be implanted with track tags and a reader is able to analyze the signals received from the tag. This chapter introduces the reader to the subject of radio frequency identification and radio frequency.

Radio-frequency Identification

Radio-frequency identification (RFID) uses electromagnetic fields to automatically identify and track tags attached to objects. The tags contain electronically stored information. Passive tags collect energy from a nearby RFID reader's interrogating radio waves. Active tags have a local power source such as a battery and may operate at hundreds of meters from the RFID reader. Unlike a barcode, the tag need not be within the line of sight of the reader, so it may be embedded in the tracked object. RFID is one method for Automatic Identification and Data Capture (AIDC).

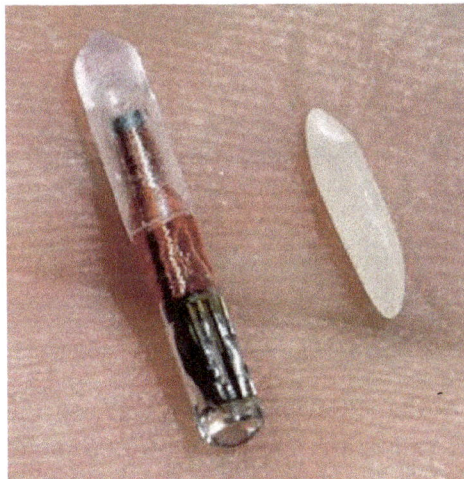

Small RFID chips, here compared to a grain of rice, are incorporated in consumer products, and implanted in pets, for identification purposes

RFID tags are used in many industries, for example, an RFID tag attached to an automobile during production can be used to track its progress through the assembly line;

RFID-tagged pharmaceuticals can be tracked through warehouses; and implanting RFID microchips in livestock and pets allows positive identification of animals.

Since RFID tags can be attached to cash, clothing, and possessions, or implanted in animals and people, the possibility of reading personally-linked information without consent has raised serious privacy concerns. These concerns resulted in standard specifications development addressing privacy and security issues. ISO/IEC 18000 and ISO/IEC 29167 use on-chip cryptography methods for untraceability, tag and reader authentication, and over-the-air privacy. ISO/IEC 20248 specifies a digital signature data structure for RFID and barcodes providing data, source and read method authenticity. This work is done within ISO/IEC JTC 1/SC 31 Automatic identification and data capture techniques.

In 2014, the world RFID market is worth US$8.89 billion, up from US$7.77 billion in 2013 and US$6.96 billion in 2012. This includes tags, readers, and software/services for RFID cards, labels, fobs, and all other form factors. The market value is expected to rise to US$18.68 billion by 2026.

History

In 1945, Léon Theremin invented an espionage tool for the Soviet Union which retransmitted incident radio waves with audio information. Sound waves vibrated a diaphragm which slightly altered the shape of the resonator, which modulated the reflected radio frequency. Even though this device was a covert listening device, not an identification tag, it is considered to be a predecessor of RFID, because it was likewise passive, being energized and activated by waves from an outside source.

FasTrak, an RFID tag used for electronic toll collection in California

Similar technology, such as the IFF transponder, was routinely used by the allies and Germany in World War II to identify aircraft as friend or foe. Transponders are still used by most powered aircraft to this day. Another early work exploring RFID is the landmark 1948 paper by Harry Stockman. Stockman predicted that "... considerable research and development work has to be done before the remaining basic problems in reflected-power communication are solved, and before the field of useful applications is explored."

Mario Cardullo's device, patented on January 23, 1973, was the first true ancestor of modern RFID, as it was a passive radio transponder with memory. The initial device was passive, powered by the interrogating signal, and was demonstrated in 1971 to the New York Port Authority and other potential users and consisted of a transponder with 16 bit memory for use as a toll device. The basic Cardullo patent covers the use of RF, sound and light as transmission media. The original business plan presented to investors in 1969 showed uses in transportation (automotive vehicle identification, automatic toll system, electronic license plate, electronic manifest, vehicle routing, vehicle performance monitoring), banking (electronic check book, electronic credit card), security (personnel identification, automatic gates, surveillance) and medical (identification, patient history).

An early demonstration of reflected power (modulated backscatter) RFID tags, both passive and semi-passive, was performed by Steven Depp, Alfred Koelle, and Robert Frayman at the Los Alamos National Laboratory in 1973. The portable system operated at 915 MHz and used 12-bit tags. This technique is used by the majority of today's UHFID and microwave RFID tags.

The first patent to be associated with the abbreviation RFID was granted to Charles Walton in 1983.

Design

Tags

A radio-frequency identification system uses tags, or labels attached to the objects to be identified. Two-way radio transmitter-receivers called interrogators or readers send a signal to the tag and read its response.

RFID tags can be either passive, active or battery-assisted passive. An active tag has an on-board battery and periodically transmits its ID signal. A battery-assisted passive (BAP) has a small battery on board and is activated when in the presence of an RFID reader. A passive tag is cheaper and smaller because it has no battery; instead, the tag uses the radio energy transmitted by the reader. However, to operate a passive tag, it must be illuminated with a power level roughly a thousand times stronger than for signal transmission. That makes a difference in interference and in exposure to radiation.

Tags may either be read-only, having a factory-assigned serial number that is used as a key into a database, or may be read/write, where object-specific data can be written into the tag by the system user. Field programmable tags may be write-once, read-multiple; "blank" tags may be written with an electronic product code by the user.

RFID tags contain at least two parts: an integrated circuit for storing and processing information, modulating and demodulating a radio-frequency (RF) signal, collecting DC power from the incident reader signal, and other specialized functions; and an antenna for receiving and transmitting the signal. The tag information is stored in a non-volatile memory. The RFID tag includes either fixed or programmable logic for processing the transmission and sensor data, respectively.

An RFID reader transmits an encoded radio signal to interrogate the tag. The RFID tag receives the message and then responds with its identification and other information. This may be only a unique tag serial number, or may be product-related information such as a stock number, lot or batch number, production date, or other specific information. Since tags have individual serial numbers, the RFID system design can discriminate among several tags that might be within the range of the RFID reader and read them simultaneously.

Readers

RFID systems can be classified by the type of tag and reader. A Passive Reader Active Tag (PRAT) system has a passive reader which only receives radio signals from active tags (battery operated, transmit only). The reception range of a PRAT system reader can be adjusted from 1–2,000 feet (0–600 m), allowing flexibility in applications such as asset protection and supervision.

An Active Reader Passive Tag (ARPT) system has an active reader, which transmits interrogator signals and also receives authentication replies from passive tags.

An Active Reader Active Tag (ARAT) system uses active tags awoken with an interrogator signal from the active reader. A variation of this system could also use a Battery-Assisted Passive (BAP) tag which acts like a passive tag but has a small battery to power the tag's return reporting signal.

Fixed readers are set up to create a specific interrogation zone which can be tightly controlled. This allows a highly defined reading area for when tags go in and out of the interrogation zone. Mobile readers may be hand-held or mounted on carts or vehicles.

Frequencies

RFID frequency bands						
Band	Regulations	Range	Data speed	ISO/IEC 18000 Section	Remarks	Approximate tag cost in volume (2006) US $
120–150 kHz (LF)	Unregulated	10 cm	Low	Part 2	Animal identification, factory data collection	$1

13.56 MHz (HF)	ISM band world-wide	10 cm–1 m	Low to moderate	Part 3	Smart cards (ISO/IEC 15693, ISO/IEC 14443 A,B). Non fully ISO compatible memory cards (Mifare Classic, iCLASS, Legic, Felica ...). Micro processor ISO compatible cards (Desfire EV1, Seos)	$0.50 to $5
433 MHz (UHF)	Short Range Devices	1–100 m	Moderate	Part 7	Defense applications, with active tags	$5
865-868 MHz (Europe) 902-928 MHz (North America) UHF	ISM band	1–12 m	Moderate to high	Part 6	EAN, various standards	$0.15 (passive tags)
2450-5800 MHz (microwave)	ISM band	1–2 m	High	Part 4	802.11 WLAN, Bluetooth standards	$25 (active tags)
3.1–10 GHz (microwave)	Ultra wide band	to 200 m	High	Not Defined	requires semi-active or active tags	$5 projected

Signaling

Signaling between the reader and the tag is done in several different incompatible ways, depending on the frequency band used by the tag. Tags operating on LF and HF bands are, in terms of radio wavelength, very close to the reader antenna because they are only a small percentage of a wavelength away. In this near field region, the tag is closely coupled electrically with the transmitter in the reader. The tag can modulate the field produced by the reader by changing the electrical loading the tag represents. By switching between lower and higher relative loads, the tag produces a change that the reader can detect. At UHF and higher frequencies, the tag is more than one radio wavelength away from the reader, requiring a different approach. The tag can backscatter a signal. Active tags may contain functionally separated transmitters and receivers, and the tag need not respond on a frequency related to the reader's interrogation signal.

An Electronic Product Code (EPC) is one common type of data stored in a tag. When written into the tag by an RFID printer, the tag contains a 96-bit string of data. The first eight bits are a header which identifies the version of the protocol. The next 28 bits identify the organization that manages the data for this tag; the organization number is assigned by the EPCGlobal consortium. The next 24 bits are an object class, identifying the kind of product; the last 36 bits are a unique serial number for a particular tag.

These last two fields are set by the organization that issued the tag. Rather like a URL, the total electronic product code number can be used as a key into a global database to uniquely identify a particular product.

Often more than one tag will respond to a tag reader, for example, many individual products with tags may be shipped in a common box or on a common pallet. Collision detection is important to allow reading of data. Two different types of protocols are used to "singulate" a particular tag, allowing its data to be read in the midst of many similar tags. In a slotted Aloha system, the reader broadcasts an initialization command and a parameter that the tags individually use to pseudo-randomly delay their responses. When using an "adaptive binary tree" protocol, the reader sends an initialization symbol and then transmits one bit of ID data at a time; only tags with matching bits respond, and eventually only one tag matches the complete ID string.

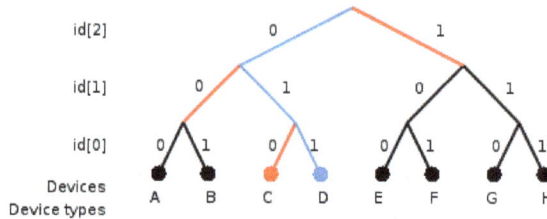

An example of a binary tree method of identifying an RFID tag

Both methods have drawbacks when used with many tags or with multiple overlapping readers. Bulk reading is a strategy for interrogating multiple tags at the same time, but lacks sufficient precision for inventory control.

Miniaturization

RFIDs are easy to conceal or incorporate in other items. For example, in 2009 researchers at Bristol University successfully glued RFID micro-transponders to live ants in order to study their behavior. This trend towards increasingly miniaturized RFIDs is likely to continue as technology advances.

Hitachi holds the record for the smallest RFID chip, at 0.05 mm × 0.05 mm. This is 1/64th the size of the previous record holder, the mu-chip. Manufacture is enabled by using the silicon-on-insulator (SOI) process. These dust-sized chips can store 38-digit numbers using 128-bit Read Only Memory (ROM). A major challenge is the attachment of antennas, thus limiting read range to only millimeters.

Uses

The RFID tag can be affixed to an object and used to track and manage inventory, assets, people, etc. For example, it can be affixed to cars, computer equipment, books, mobile phones, etc.

RFID offers advantages over manual systems or use of bar codes. The tag can be read if passed near a reader, even if it is covered by the object or not visible. The tag can be read inside a case, carton, box or other container, and unlike barcodes, RFID tags can be read hundreds at a time. Bar codes can only be read one at a time using current devices.

In 2011, the cost of passive tags started at US$0.09 each; special tags, meant to be mounted on metal or withstand gamma sterilization, can go up to US$5. Active tags for tracking containers, medical assets, or monitoring environmental conditions in data centers start at US$50 and can go up over US$100 each. Battery-Assisted Passive (BAP) tags are in the US$3–10 range and also have sensor capability like temperature and humidity.

RFID can be used in a variety of applications, such as:

Electronic Lock with RFID Card System, ANSI

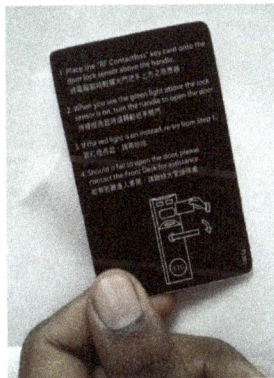

Electronic key for RFID based lock system

- Access management

- Tracking of goods

- Tracking of persons and animals

- Toll collection and contactless payment

- Machine readable travel documents

- Smartdust (for massively distributed sensor networks)

- Tracking sports memorabilia to verify authenticity

- Airport baggage tracking logistics

- Timing sporting events

In 2010 three factors drove a significant increase in RFID usage: decreased cost of equipment and tags, increased performance to a reliability of 99.9% and a stable international standard around UHF passive RFID. The adoption of these standards were driven by EPCglobal, a joint venture between GS1 and GS1 US, which were responsible for driving global adoption of the barcode in the 1970s and 1980s. The EPCglobal Network was developed by the Auto-ID Center.

Commerce

RFID provides a way for organizations to identify and manage tools and equipment (asset tracking) , without manual data entry. RFID is being adopted for item level tagging in retail stores. This provides electronic article surveillance (EAS), and a self checkout process for consumers. Automatic identification with RFID can be used for inventory systems. Manufactured products such as automobiles or garments can be tracked through the factory and through shipping to the customer.

An EPC RFID tag used by Wal-Mart.

Casinos can use RFID to authenticate poker chips, and can selectively invalidate any chips known to be stolen.

Sewn-in RFID label in garment manufactured by the French sports supplier Decathlon. Front, back, and transparency scan.

Wal-Mart and the United States Department of Defense have published requirements that their vendors place RFID tags on all shipments to improve supply chain management.

Access Control

Rfid Tags are widely used in identification badges, replacing earlier magnetic stripe cards. These badges need only be held within a certain distance of the reader to authenticate the holder. Tags can also be placed on vehicles, which can be read at a distance, to allow entrance to controlled areas without having to stop the vehicle and present a card or enter an access code.

RFID antenna for vehicular access control

Advertising

In 2010 Vail Resorts began using UHF Passive RFID tags in ski passes. Facebook is using RFID cards at most of their live events to allow guests to automatically capture and post photos. The automotive brands have adopted RFID for social media product placement more quickly than other industries. Mercedes was an early adopter in 2011 at the PGA Golf Championships, and by the 2013 Geneva Motor Show many of the larger brands were using RFID for social media marketing.

Promotion Tracking

To prevent retailers diverting products, manufacturers are exploring the use of RFID tags on promoted merchandise so that they can track exactly which product has sold through the supply chain at fully discounted prices.

Transportation and Logistics

Yard management, shipping and freight and distribution centers use RFID tracking. In the railroad industry, RFID tags mounted on locomotives and rolling stock identify the owner, identification number and type of equipment and its characteristics. This can be used with a database to identify the lading, origin, destination, etc. of the commodities being carried.

In commercial aviation, RFID is used to support maintenance on commercial aircraft. RFID tags are used to identify baggage and cargo at several airports and airlines.

Some countries are using RFID for vehicle registration and enforcement. RFID can help detect and retrieve stolen cars.

Intelligent Transportation Systems

RFID is used in intelligent transportation systems. In New York City, RFID readers are deployed at intersections to track E-ZPass tags as a means for monitoring the traffic flow. The data is fed through the broadband wireless infrastructure to the traffic management center to be used in adaptive traffic control of the traffic lights.

RFID E-ZPass reader attached to the pole and antenna (right) used in traffic monitoring in New York City

Hose Stations and Conveyance of Fluids

The RFID antenna in a permanently installed coupling half (fixed part) unmistakably identifies the RFID transponder placed in the other coupling half (free part) after completed coupling. When connected the transponder of the free part transmits all important information contactless to the fixed part. The coupling's location can be clearly identified by the RFID transponder coding. The control is enabled to automatically start subsequent process steps.

Public Transport

RFID cards are used for access control to public transport.

In London travellers use Oyster Cards on the tube, buses and ferries. It identifies the traveller at each turnstile and so the system can calculate the fare.

In the Chicago area, riders use the open standard Ventra card to board CTA buses and trains, along with PACE buses.

In Ontario, Canada, riders in the GTA and Ottawa Area use the Presto card to board trains, buses and street cars across multiple different transit companies.

Infrastructure Management and Protection

At least one company has introduced RFID to identify and locate underground infrastructure assets such as gas pipelines, sewer lines, electrical cables, communication cables, etc.

Passports

The first RFID passports ("E-passport") were issued by Malaysia in 1998. In addition to information also contained on the visual data page of the passport, Malaysian e-passports record the travel history (time, date, and place) of entries and exits from the country.

Other countries that insert RFID in passports include Norway (2005), Japan (March 1, 2006), most EU countries (around 2006), Australia, Hong Kong, the United States (2007), India (June 2008), Serbia (July 2008), Republic of Korea (August 2008), Taiwan (December 2008), Albania (January 2009), The Philippines (August 2009), Republic of Macedonia (2010), and Canada (2013).

Standards for RFID passports are determined by the International Civil Aviation Organization (ICAO), and are contained in ICAO Document 9303, Part 1, Volumes 1 and 2 (6th edition, 2006). ICAO refers to the ISO/IEC 14443 RFID chips in e-passports as "contactless integrated circuits". ICAO standards provide for e-passports to be identifiable by a standard e-passport logo on the front cover.

Since 2006, RFID tags included in new United States passports will store the same information that is printed within the passport, and include a digital picture of the owner. The United States Department of State initially stated the chips could only be read from a distance of 10 centimetres (3.9 in), but after widespread criticism and a clear demonstration that special equipment can read the test passports from 10 metres (33 ft) away, the passports were designed to incorporate a thin metal lining to make it more difficult for unauthorized readers to "skim" information when the passport is closed. The department will also implement Basic Access Control (BAC), which functions as a Personal Identification Number (PIN) in the form of characters printed on the passport data page. Before a passport's tag can be read, this PIN must be entered into an RFID reader. The BAC also enables the encryption of any communication between the chip and interrogator. As noted in the section below on security, there are many situations in which these protections have been shown to be insufficient, and passports have been cloned based on scans of them while they were being delivered in the mail.

Transportation Payments

In many countries, RFID tags can be used to pay for mass transit fares on bus, trains, or subways, or to collect tolls on highways.

Some bike lockers are operated with RFID cards assigned to individual users. A pre-paid card is required to open or enter a facility or locker and is used to track and charge based on how long the bike is parked.

The Zipcar car-sharing service uses RFID cards for locking and unlocking cars and for member identification.

In Singapore, RFID replaces paper Season Parking Ticket (SPT).

Animal Identification

RFID tags for animals represent one of the oldest uses of RFID. Originally meant for large ranches and rough terrain, since the outbreak of mad-cow disease, RFID has become crucial in animal identification management. An implantable RFID tag or transponder can also be used for animal identification. The transponders are more well known as passive RFID, or "chips" on animals. The Canadian Cattle Identification Agency began using RFID tags as a replacement for barcode tags. Currently CCIA tags are used in Wisconsin and by United States farmers on a voluntary basis. The USDA is currently developing its own program.

RFID tags are required for all cattle, sheep and goats sold in Australia.

Human Identification

Implantable RFID chips designed for animal tagging are now being used in humans. An early experiment with RFID implants was conducted by British professor of cybernetics Kevin Warwick, who implanted a chip in his arm in 1998. In 2004 Conrad Chase offered implanted chips in his night clubs in Barcelona and Rotterdam to identify their VIP customers, who in turn use it to pay for drinks.

The Food and Drug Administration in the United States has approved the use of RFID chips in humans. Some business establishments give customers the option of using an RFID-based tab to pay for service, such as the Baja Beach nightclub in Barcelona. This has provoked concerns into privacy of individuals as they can potentially be tracked wherever they go by an identifier unique to them. Some are concerned this could lead to abuse by an authoritarian government, to removal of freedoms, and to the emergence of an "ultimate panopticon", a society where all citizens behave in a socially accepted manner because others might be watching.

On July 22, 2006, Reuters reported that two hackers, Newitz and Westhues, at a conference in New York City showed that they could clone the RFID signal from a human implanted RFID chip, showing that the chip is not hack-proof as was previously claimed. Privacy advocates have protested against implantable RFID chips, warning of potential abuse. There is much controversy regarding human applications of this technology, and many conspiracy theories abound in relation to human applications, especially one of which is referred to as "The Mark of the Beast" in some religious circles.

Institutions

Hospitals and Healthcare

In Healthcare, there is a need for increased visibility, efficiency, and gathering of data around relevant interactions. RFID tracking solutions are able to help healthcare facilities manage mobile medical equipment, improve patient workflow, monitor environmental conditions, and protect patients, staff and visitors from infection or other hazards.

Adoption of RFID in the medical industry has been widespread and very effective. Hospitals are among the first users to combine both active and passive RFID. Many successful deployments in the healthcare industry have been cited where active technology tracks high-value, or frequently moved items, where passive technology tracks smaller, lower cost items that only need room-level identification. For example, medical facility rooms can collect data from transmissions of RFID badges worn by patients and employees, as well as from tags assigned to facility assets, such as mobile medical devices. The U.S. Department of Veterans Affairs (VA) recently announced plans to deploy RFID in hospitals across America to improve care and reduce costs.

A physical RFID tag may be incorporated with browser-based software to increase its efficacy. This software allows for different groups or specific hospital staff, nurses, and patients to see real-time data relevant to each piece of tracked equipment or personnel. Real-time data is stored and archived to make use of historical reporting functionality and to prove compliance with various industry regulations. This combination of RFID real-time locating system hardware and software provides a powerful data collection tool for facilities seeking to improve operational efficiency and reduce costs.

The trend is toward using ISO 18000-6c as the tag of choice and combining an active tagging system that relies on existing 802.11X wireless infrastructure for active tags.

Since 2004 a number of U.S. hospitals have begun implanting patients with RFID tags and using RFID systems, usually for workflow and inventory management. The use of RFID to prevent mixups between sperm and ova in IVF clinics is also being considered.

In October 2004, the FDA approved USA's first RFID chips that can be implanted in humans. The 134 kHz RFID chips, from VeriChip Corp. can incorporate personal medical information and could save lives and limit injuries from errors in medical treatments, according to the company. Anti-RFID activists Katherine Albrecht and Liz McIntyre discovered an FDA Warning Letter that spelled out health risks. According to the FDA, these include "adverse tissue reaction", "migration of the implanted transponder", "failure of implanted transponder", "electrical hazards" and "magnetic resonance imaging [MRI] incompatibility."

Libraries

Libraries have used RFID to replace the barcodes on library items. The tag can contain identifying information or may just be a key into a database. An RFID system may replace or supplement bar codes and may offer another method of inventory management and self-service checkout by patrons. It can also act as a security device, taking the place of the more traditional electromagnetic security strip.

RFID tags used in libraries: square book tag, round CD/DVD tag and rectangular VHS tag

It is estimated that over 30 million library items worldwide now contain RFID tags, including some in the Vatican Library in Rome.

Since RFID tags can be read through an item, there is no need to open a book cover or DVD case to scan an item, and a stack of books can be read simultaneously. Book tags can be read while books are in motion on a conveyor belt, which reduces staff time. This can all be done by the borrowers themselves, reducing the need for library staff assistance. With portable readers, inventories could be done on a whole shelf of materials within seconds. However, as of 2008 this technology remains too costly for many smaller libraries, and the conversion period has been estimated at 11 months for an average-size library. A 2004 Dutch estimate was that a library which lends 100,000 books per year should plan on a cost of €50,000 (borrow- and return-stations: 12,500 each, detection porches 10,000 each; tags 0.36 each). RFID taking a large burden off staff could also mean that fewer staff will be needed, resulting in some of them getting laid off, but that has so far not happened in North America where recent surveys have not returned a single library that cut staff because of adding RFID. In fact, library budgets are being reduced for personnel and increased for infrastructure, making it necessary for libraries to add automation to compensate for the reduced staff size. Also, the tasks that RFID takes over are largely not the primary tasks of librarians. A finding in the Netherlands is that borrowers are pleased with the fact that staff are now more available for answering questions.

Privacy concerns have been raised surrounding library use of RFID. Because some RFID tags can be read from up to 100 metres (330 ft), there is some concern over whether sensitive information could be collected from an unwilling source. However, library RFID tags do not contain any patron information, and the tags used in the majority of libraries use a frequency only readable from approximately 10 feet (3.0 m). Further, another non-library agency could potentially record the RFID tags of every person leaving the library without the library administrator's knowledge or consent. One simple option is to let the book transmit a code that has meaning only in conjunction with the library's database. Another possible enhancement would be to give each book a new code every time it is returned. In future, should readers become ubiquitous (and possibly networked), then stolen books could be traced even outside the library. Tag removal could be made difficult if the tags are so small that they fit invisibly inside a (random) page, possibly put there by the publisher.

Museums

RFID technologies are now also implemented in end-user applications in museums. An example was the custom-designed temporary research application, "eXspot," at the Exploratorium, a science museum in San Francisco, California. A visitor entering the museum received an RF Tag that could be carried as a card. The eXspot system enabled the visitor to receive information about specific exhibits. Aside from the exhibit information, the visitor could take photographs of themselves at the exhibit. It was also intended to allow the visitor to take data for later analysis. The collected information could be retrieved at home from a "personalized" website keyed to the RFID tag.

Schools and Universities

School authorities in the Japanese city of Osaka are now chipping children's clothing, backpacks, and student IDs in a primary school. A school in Doncaster, England is piloting a monitoring system designed to keep tabs on pupils by tracking radio chips in their uniforms. St Charles Sixth Form College in west London, England, started September, 2008, is using an RFID card system to check in and out of the main gate, to both track attendance and prevent unauthorized entrance. Similarly, Whitcliffe Mount School in Cleckheaton, England uses RFID to track pupils and staff in and out of the building via a specially designed card. In the Philippines, some schools already use RFID in IDs for borrowing books and also gates in those particular schools have RFID ID scanners for buying items at a school shop and canteen, library and also to sign in and sign out for student and teacher's attendance.

Sports

RFID for timing races began in the early 1990s with pigeon racing, introduced by the company Deister Electronics in Germany. RFID can provide race start and end timings for individuals in large races where it is impossible to get accurate stopwatch readings for every entrant.

ChampionChip

In the race, the racers wear tags that are read by antennas placed alongside the track or on mats across the track. UHF tags provide accurate readings with specially designed antennas. Rush error, lap count errors and accidents at start time are avoided since anyone can start and finish any time without being in a batch mode.

J-Chip 8-channel receiver next to timing mat. The athlete wears a chip on a strap around his ankle. Ironman Germany 2007 in Frankfurt.

The design of chip+antenna controls the range from which it can be read. Short range compact chips are twist tied to the shoe or velcro strapped the ankle. These need to be about 400mm from the mat and so give very good temporal resolution. Alternatively a chip plus a very large (a 125mm square) antenna can be incorporated into the bib number worn on the athlete's chest at about 1.25m height.

Passive and active RFID systems are used in off-road events such as Orienteering, Enduro and Hare and Hounds racing. Riders have a transponder on their person, normally on their arm. When they complete a lap they swipe or touch the receiver which is connected to a computer and log their lap time.

RFID is being adapted by many recruitment agencies which have a PET (Physical Endurance Test) as their qualifying procedure especially in cases where the candidate volumes may run into millions (Indian Railway Recruitment Cells, Police and Power sector).

A number of ski resorts have adopted RFID tags to provide skiers hands-free access to ski lifts. Skiers do not have to take their passes out of their pockets. Ski jackets have a left pocket into which the chip+card fits. This nearly contacts the sensor unit on the left of the turnstile as the skier pushes through to the lift. These systems were based on high frequency (HF) at 13.56 megahertz. The bulk of ski areas in Europe, from Verbier to Chamonix use these systems.

The NFL in the United States equips players with RFID chips that measures speed, distance and direction traveled by each player in real-time. Currently cameras stay focused on the quarterback, however, numerous plays are happening simultaneously on the field. The RFID chip will provide new insight into these simultaneous plays. The chip triangulates the player's position within six inches and will be used to digitally broadcast replays. The RFID chip will make individual player information accessible to the public. The data will be available via the NFL 2015 app. The RFID chips are manufactured by Zebra Technologies. Zebra Technologies tested the RFID chip in 18 stadiums last year to track vector data.

Complement to Barcode

RFID tags are often a complement, but not a substitute, for UPC or EAN barcodes. They may never completely replace barcodes, due in part to their higher cost and the advantage of multiple data sources on the same object. Also, unlike RFID labels, barcodes can be generated and distributed electronically, e.g. via e-mail or mobile phone, for printing or display by the recipient. An example is airline boarding passes. The new EPC, along with several other schemes, is widely available at reasonable cost.

The storage of data associated with tracking items will require many terabytes. Filtering and categorizing RFID data is needed to create useful information. It is likely that goods will be tracked by the pallet using RFID tags, and at package level with Universal Product Code (UPC) or EAN from unique barcodes.

The unique identity is a mandatory requirement for RFID tags, despite special choice of the numbering scheme. RFID tag data capacity is large enough that each individual tag will have a unique code, while current bar codes are limited to a single type code for a particular product. The uniqueness of RFID tags means that a product may be tracked as it moves from location to location, finally ending up in the consumer's hands. This may help to combat theft and other forms of product loss. The tracing of products is an important feature that gets well supported with RFID tags containing a unique identity of the tag and also the serial number of the object. This may help companies cope with quality deficiencies and resulting recall campaigns, but also contributes to concern about tracking and profiling of consumers after the sale.

Telemetry

Active RFID tags also have the potential to function as low-cost remote sensors that broadcast telemetry back to a base station. Applications of tagometry data could include sensing of road conditions by implanted beacons, weather reports, and noise level monitoring.

Passive RFID tags can also report sensor data. For example, the Wireless Identification and Sensing Platform is a passive tag that reports temperature, acceleration and capacitance to commercial Gen2 RFID readers.

It is possible that active or battery-assisted passive (BAP) RFID tags, could broadcast a signal to an in-store receiver to determine whether the RFID tag (product) is in the store.

Optical RFID

Optical RFID (aka OPID) is an alternative to RFID that is based on optical readers. Applications for optical RFID tags may be found in future supply chain scenarios. The main advantage in comparison to traditional RFID tags is their low price and the usually employed offline preaggregation of data to the class level.

Unlike most other RFID chips (which use radio frequencies of 0.125–0.1342, 0.140–0.1485, 13.56, and 868–928 MHz), optical RFID operates in the electromagnetic spectrum between the frequencies of 333 THz (3.33×10^{14} hertz, 900 nm) and 380 THz (788 nm) and 750 THz (400 nm). The tag information is communicated to the reader by reflecting the read request. Parts of the incoming signal are filtered by the tag in a well-defined way as it is sent back to the reader. On the reader's side, the tag data can be deduced by analysing the pattern used for filtering. As an alternative to reflection mode, active circuits can be used, replacing awkward RFID antennae with photovoltaic components and IR-LEDs on the ICs. One of the earliest examples of Optical RFID is RFIG: Radio Frequency Identity and Geometry, by Ramesh Raskar, Paul Dietz, Paul Beardsley and colleagues. This combines an optical tag with a RF tag to provide ID as well as geometric operations, such as location, pose, motion, orientation and change detection.

Regarding privacy, optical RFID provides much more protection against abuse than RFID based on common electromagnetic waves. This is mainly because line-of-sight is required for malicious read out. Such an attack can easily be prevented with low cost optical RFID sight blockers. Nevertheless, if needed some penetration of solids and liquids can be achieved e.g. with near-IR wavelengths.

Regulation and Standardization

A number of organizations have set standards for RFID, including the International Organization for Standardization (ISO), the International Electrotechnical Commission (IEC), ASTM International, the DASH7 Alliance and EPCglobal.

There are also several specific industries that have set guidelines. These industries include the Financial Services Technology Consortium (FSTC) which has set a standard for tracking IT Assets with RFID, the Computer Technology Industry Association CompTIA which has set a standard for certifying RFID engineers, and the International

Airlines Transport Association IATA which has set tagging guidelines for luggage in airports.

In principle, every country can set its own rules for frequency allocation for RFID tags, and not all radio bands are available in all countries. These frequencies are known as the ISM bands (Industrial Scientific and Medical bands). The return signal of the tag may still cause interference for other radio users.

- Low-frequency (LF: 125–134.2 kHz and 140–148.5 kHz) (LowFID) tags and high-frequency (HF: 13.56 MHz) (HighFID) tags can be used globally without a license.

- Ultra-high-frequency (UHF: 865–928 MHz) (Ultra-HighFID or UHFID) tags cannot be used globally as there is no single global standard and regulations differ from country to country.

In North America, UHF can be used unlicensed for 902–928 MHz ($\pm$13 MHz from the 915 MHz center frequency), but restrictions exist for transmission power. In Europe, RFID and other low-power radio applications are regulated by ETSI recommendations EN 300 220 and EN 302 208, and ERO recommendation 70 03, allowing RFID operation with somewhat complex band restrictions from 865–868 MHz. Readers are required to monitor a channel before transmitting ("Listen Before Talk"); this requirement has led to some restrictions on performance, the resolution of which is a subject of current research. The North American UHF standard is not accepted in France as it interferes with its military bands. On July 25, 2012, Japan changed its UHF band to 920 MHz, more closely matching the United States' 915 MHz band.

In some countries, a site license is needed, which needs to be applied for at the local authorities, and can be revoked.

According to an overview assembled by GS1, as of 31 October 2014, regulations are in place in 78 countries representing ca. 96.5% of the world's GDP, and work on regulations is in progress in 3 countries representing circa 1% of the world's GDP.

Standards that have been made regarding RFID include:

- ISO 14223 – Radiofrequency [sic] identification of animals – Advanced transponders

- ISO/IEC 14443: This standard is a popular HF (13.56 MHz) standard for HighFIDs which is being used as the basis of RFID-enabled passports under ICAO 9303. The Near Field Communication standard that lets mobile devices act as RFID readers/transponders is also based on ISO/IEC 14443.

- ISO/IEC 15693: This is also a popular HF (13.56 MHz) standard for HighFIDs widely used for non-contact smart payment and credit cards.

- ISO/IEC 18000: Information technology—Radio frequency identification for item management:

  - Part 1: Reference architecture and definition of parameters to be standardized

  - Part 2: Parameters for air interface communications below 135 kHz

  - Part 3: Parameters for air interface communications at 13.56 MHz

  - Part 4: Parameters for air interface communications at 2.45 GHz

  - Part 6: Parameters for air interface communications at 860–960 MHz

  - Part 7: Parameters for active air interface communications at 433 MHz

- ISO/IEC 18092 Information technology—Telecommunications and information exchange between systems—Near Field Communication—Interface and Protocol (NFCIP-1)

- ISO 18185: This is the industry standard for electronic seals or "e-seals" for tracking cargo containers using the 433 MHz and 2.4 GHz frequencies.

- ISO/IEC 21481 Information technology—Telecommunications and information exchange between systems—Near Field Communication Interface and Protocol -2 (NFCIP-2)

- ASTM D7434, Standard Test Method for Determining the Performance of Passive Radio Frequency Identification (RFID) Transponders on Palletized or Unitized Loads

- ASTM D7435, Standard Test Method for Determining the Performance of Passive Radio Frequency Identification (RFID) Transponders on Loaded Containers

- ASTM D7580, Standard Test Method for Rotary Stretch Wrapper Method for Determining the Readability of Passive RFID Transponders on Homogenous Palletized or Unitized Loads

- ISO 28560-2 : specifies encoding standards and data model to be used within libraries.

In order to ensure global interoperability of products, several organizations have set up additional standards for RFID testing. These standards include conformance, performance and interoperability tests.

Groups concerned with standardization are:

- DASH7 Alliance – an international industry group formed in 2009 to promote standards and interoperability among extensions to ISO/IEC 18000-7 technologies

- EPCglobal – this is the standardization framework that is most likely to undergo international standardisation according to ISO rules as with all sound standards in the world, unless residing with limited scope, as customs regulations, air-traffic regulations and others. Currently the big distributors and governmental customers are pushing EPC heavily as a standard well-accepted in their community, but not yet regarded as for salvation to the rest of the world.

EPC Gen2

EPC Gen2 is short for EPCglobal UHF Class 1 Generation 2.

EPCglobal, a joint venture between GS1 and GS1 US, is working on international standards for the use of mostly passive RFID and the Electronic Product Code (EPC) in the identification of many items in the supply chain for companies worldwide.

One of the missions of EPCglobal was to simplify the Babel of protocols prevalent in the RFID world in the 1990s. Two tag air interfaces (the protocol for exchanging information between a tag and a reader) were defined (but not ratified) by EPCglobal prior to 2003. These protocols, commonly known as Class 0 and Class 1, saw significant commercial implementation in 2002–2005.

In 2004, the Hardware Action Group created a new protocol, the Class 1 Generation 2 interface, which addressed a number of problems that had been experienced with Class 0 and Class 1 tags. The EPC Gen2 standard was approved in December 2004. This was approved after a contention from Intermec that the standard may infringe a number of their RFID-related patents. It was decided that the standard itself does not infringe their patents, making the standard royalty free. The EPC Gen2 standard was adopted with minor modifications as ISO 18000-6C in 2006.

In 2007, the lowest cost of Gen2 EPC inlay was offered by the now-defunct company SmartCode, at a price of $0.05 apiece in volumes of 100 million or more. Nevertheless, further conversion (including additional label stock or encapsulation processing/insertion and freight costs to a given facility or DC) and of the inlays into usable RFID labels and the design of current Gen 2 protocol standard will increase the total end-cost, especially with the added security feature extensions for RFID Supply Chain item-level tagging.

Problems and Concerns

Data Flooding

Not every successful reading of a tag (an observation) is useful for business purposes. A large amount of data may be generated that is not useful for managing inventory or other applications. For example, a customer moving a product from one shelf to another, or a pallet load of articles that passes several readers while being moved in a warehouse, are events that do not produce data that is meaningful to an inventory control system.

Event filtering is required to reduce this data inflow to a meaningful depiction of moving goods passing a threshold. Various concepts have been designed, mainly offered as middleware performing the filtering from noisy and redundant raw data to significant processed data.

Global Standardization

The frequencies used for UHF RFID in the USA are currently incompatible with those of Europe or Japan. Furthermore, no emerging standard has yet become as universal as the barcode. To address international trade concerns, it is necessary to use a tag that is operational within all of the international frequency domains.

Security Concerns

Retailers such as Walmart, which already heavily use RFID for inventory purposes, also use RFID as an anti-employee-theft and anti-shoplifting technology. If a product with an active RFID tag passes the exit-scanners at a Walmart outlet, not only does it set off an alarm, but it also tells security personnel exactly what product to look for in the shopper's cart.

A primary RFID security concern is the illicit tracking of RFID tags. Tags, which are world-readable, pose a risk to both personal location privacy and corporate/military security. Such concerns have been raised with respect to the United States Department of Defense's recent adoption of RFID tags for supply chain management. More generally, privacy organizations have expressed concerns in the context of ongoing efforts to embed electronic product code (EPC) RFID tags in consumer products. This is mostly as result of the fact that RFID tags can be read, and legitimate transactions with readers can be eavesdropped, from non-trivial distances. RFID used in access control, payment and eID (e-passport) systems operate at a shorter range than EPC RFID systems but are also vulnerable to skimming and eavesdropping, albeit at shorter distance.

A second method of prevention is by using cryptography. Rolling codes and challenge-response authentication (CRA) are commonly used to foil monitor-repetition of the messages between the tag and reader; as any messages that have been recorded would prove to be unsuccessful on repeat transmission. Rolling codes rely upon the tag's id being changed after each interrogation, while CRA uses software to ask for a cryptographically coded response from the tag. The protocols used during CRA can be symmetric, or may use public key cryptography.

Security concerns exist in regard to privacy over the unauthorized reading of RFID tags. Unauthorized readers can potentially use RFID information to identify or track packages, consumers, carriers, or the contents of a package. Several prototype systems are being developed to combat unauthorized reading, including RFID signal interruption, as well as the possibility of legislation, and 700 scientific papers have been published on this matter since 2002. There are also concerns that the database structure of Object Naming Services may be susceptible to infiltration, similar to denial-of-service attacks, after the EPCglobal Network ONS root servers were shown to be vulnerable.

Exploitation

Ars Technica reported in March 2006 an RFID buffer overflow bug that could infect airport terminal RFID databases for baggage, and also passport databases to obtain confidential information on the passport holder.

Passports

In an effort to standardize and make it easier to process passports, several countries have implemented RFID in passports, despite security and privacy issues. The encryption on UK chips was broken in under 48 hours. Since that incident, further efforts have allowed researchers to clone passport data while the passport is being mailed to its owner. Where a criminal used to need to secretly open and then reseal the envelope, now it can be done without detection, adding some degree of insecurity to the passport system.

Shielding

In an effort to prevent the passive "skimming" of RFID-enabled cards or passports, the U.S. General Services Administration (GSA) issued a set of test procedures for evaluating electromagnetically opaque sleeves. For shielding products to be in compliance with FIPS-201 guidelines, they must meet or exceed this published standard. Shielding products currently evaluated as FIPS-201 compliant are listed on the website of the U.S. CIO's FIPS-201 Evaluation Program. The United States government requires that when new ID cards are issued, they must be delivered with an approved shielding sleeve or holder.

There are contradicting opinions as to whether aluminum can prevent reading of RFID chips. Some people claim that aluminum shielding, essentially creating a Faraday cage, does work. Others claim that simply wrapping an RFID card in aluminum foil only makes transmission more difficult and is not completely effective at preventing it.

Shielding effectiveness depends on the frequency being used. Low-frequency LowFID tags, like those used in implantable devices for humans and pets, are relatively resistant to shielding though thick metal foil will prevent most reads. High frequency HighFID tags (13.56 MHz—smart cards and access badges) are sensitive to shielding and are difficult to read when within a few centimetres of a metal surface. UHF Ultra-HighFID tags (pallets and cartons) are difficult to read when placed within a few millimetres of a metal surface, although their read range is actually increased when they are spaced

2–4 cm from a metal surface due to positive reinforcement of the reflected wave and the incident wave at the tag.

Controversies

Logo of the anti-RFID campaign by German privacy group digitalcourage (formerly FoeBuD).

Privacy

The use of RFID has engendered considerable controversy and even product boycotts by consumer privacy advocates. Consumer privacy experts Katherine Albrecht and Liz McIntyre are two prominent critics of the "spychip" technology. The two main privacy concerns regarding RFID are:

- Since the owner of an item will not necessarily be aware of the presence of an RFID tag and the tag can be read at a distance without the knowledge of the individual, it becomes possible to gather sensitive data about an individual without consent.

- If a tagged item is paid for by credit card or in conjunction with use of a loyalty card, then it would be possible to indirectly deduce the identity of the purchaser by reading the globally unique ID of that item (contained in the RFID tag). This is only true if the person doing the watching also had access to the loyalty card data and the credit card data, and the person with the equipment knows where you are going to be.

Most concerns revolve around the fact that RFID tags affixed to products remain functional even after the products have been purchased and taken home and thus can be used for surveillance and other purposes unrelated to their supply chain inventory functions.

The RFID Network argued that these fears are unfounded in the first episode of their syndicated cable TV series by letting RF engineers demonstrate how RFID works. They provided images of RF engineers driving an RFID-enabled van around a building and trying to take an inventory of items inside. They discussed satellite tracking of a passive RFID tag, which is surprising since the maximum range is under 200m.

The concerns raised by the above may be addressed in part by use of the Clipped Tag. The Clipped Tag is an RFID tag designed to increase consumer privacy. The Clipped Tag has been suggested by IBM researchers Paul Moskowitz and Guenter Karjoth. After the point of sale, a consumer may tear off a portion of the tag. This allows the transformation of a long-range tag into a proximity tag that still may be read, but only at short range – less than a few inches or centimeters. The modification of the tag may be confirmed visually. The tag may still be used later for returns, recalls, or recycling.

However, read range is both a function of the reader and the tag itself. Improvements in technology may increase read ranges for tags. Tags may be read at longer ranges than they are designed for by increasing reader power. The limit on read distance then becomes the signal-to-noise ratio of the signal reflected from the tag back to the reader. Researchers at two security conferences have demonstrated that passive Ultra-HighFID tags normally read at ranges of up to 30 feet, can be read at ranges of 50 to 69 feet using suitable equipment.

In January 2004 privacy advocates from CASPIAN and the German privacy group FoeBuD were invited to the METRO Future Store in Germany, where an RFID pilot project was implemented. It was uncovered by accident that METRO "Payback" customer loyalty cards contained RFID tags with customer IDs, a fact that was disclosed neither to customers receiving the cards, nor to this group of privacy advocates. This happened despite assurances by METRO that no customer identification data was tracked and all RFID usage was clearly disclosed.

During the UN World Summit on the Information Society (WSIS) between the 16th to 18 November 2005, founder of the free software movement, Richard Stallman, protested the use of RFID security cards by covering his card with aluminum foil.

In 2004–2005 the Federal Trade Commission Staff conducted a workshop and review of RFID privacy concerns and issued a report recommending best practices.

RFID was one of the main topics of 2006 Chaos Communication Congress (organized by the Chaos Computer Club in Berlin) and triggered a big press debate. Topics included: electronic passports, Mifare cryptography and the tickets for the FIFA World Cup 2006. Talks showed how the first real world mass application of RFID at the 2006 FIFA Soccer World Cup worked. Group monochrom staged a special 'Hack RFID' song.

Government Control

Some individuals have grown to fear the loss of rights due to RFID human implantation.

By early 2007, Chris Paget of San Francisco, California, showed that RFID information can be pulled from individuals by using only $250 worth of equipment. This supports the claim that with the information captured, it would be relatively simple to make counterfeit passports.

According to ZDNet, critics believe that RFID will lead to tracking individuals' every movement and will be an invasion of privacy. In the book SpyChips: How Major Corporations and Government Plan to Track Your Every Move by Katherine Albrecht and Liz McIntyre, one is encouraged to "imagine a world of no privacy. Where your every purchase is monitored and recorded in a database and your every belonging is numbered. Where someone many states away or perhaps in another country has a record of everything you have ever bought. What's more, they can be tracked and monitored remotely".

Deliberate Destruction in Clothing and Other Items

According to an RSA laboratories FAQ, RFID tags can be destroyed by a standard microwave oven; however some types of RFID tags, particularly those constructed to radiate using large metallic antennas (in particular RF tags and EPC tags), may catch fire if subjected to this process for too long (as would any metallic item inside a microwave oven). This simple method cannot safely be used to deactivate RFID features in electronic devices, or those implanted in living tissue, because of the risk of damage to the "host". However the time required is extremely short (a second or two of radiation) and the method works in many other non-electronic and inanimate items, long before heat or fire become of concern.

Some RFID tags implement a "kill command" mechanism for permanently and irreversibly disabling them. This mechanism can be applied if the chip itself is trusted or the mechanism is known by the person that wants to "kill" the tag.

UHF RFID tags that comply with the EPC2 Gen 2 Class 1 standard usually support this mechanism, while protecting the chip from being killed with a password. Guessing or cracking this needed 32-bit password for killing a tag would not be difficult for a determined attacker.

Radio Frequency

Radio frequency (RF) is any of the electromagnetic wave frequencies that lie in the range extending from around 3 kHz to 300 GHz, which include those frequencies used for communications or radar signals. RF usually refers to electrical rather than mechanical oscillations. However, mechanical RF systems do exist.

Although radio frequency is a rate of oscillation, the term "radio frequency" or its abbreviation "RF" are used as a synonym for radio – i.e., to describe the use of wireless communication, as opposed to communication via electric wires. Examples include:

- Radio-frequency identification

- ISO/IEC 14443-2 Radio frequency power and signal interface

Special Properties of RF Current

Electric currents that oscillate at radio frequencies have special properties not shared by direct current or alternating current of lower frequencies.

- The energy in an RF current can radiate off a conductor into space as electromagnetic waves (radio waves); this is the basis of radio technology.

- RF current does not penetrate deeply into electrical conductors but tends to flow along their surfaces; this is known as the skin effect. For this reason, when the human body comes in contact with high power RF currents it can cause superficial but serious burns called RF burns (Radiation burns).

- RF currents applied to the body often do not cause the painful sensation of electric shock as do lower frequency currents. This is because the current changes direction too quickly to trigger depolarization of nerve membranes.

- RF current can easily ionize air, creating a conductive path through it. This property is exploited by "high frequency" units used in electric arc welding, which use currents at higher frequencies than power distribution uses.

- Another property is the ability to appear to flow through paths that contain insulating material, like the dielectric insulator of a capacitor.

- When conducted by an ordinary electric cable, RF current has a tendency to reflect from discontinuities in the cable such as connectors and travel back down the cable toward the source, causing a condition called standing waves. Therefore, RF current must be carried by specialized types of cable called transmission line.

Radio Communication

To receive radio signals an antenna must be used. However, since the antenna will pick up thousands of radio signals at a time, a radio tuner is necessary to tune into a particular frequency (or frequency range). This is typically done via a resonator – in its simplest form, a circuit with a capacitor and an inductor form a tuned circuit. The resonator amplifies oscillations within a particular frequency band, while reducing oscillations at other frequencies outside the band. Another method to isolate a particular radio frequency is by oversampling (which gets a wide range of frequencies) and picking out the frequencies of interest, as done in software defined radio.

The distance over which radio communications is useful depends significantly on things other than wavelength, such as transmitter power, receiver quality, type, size, and height of antenna, mode of transmission, noise, and interfering signals. Ground waves, tropospheric scatter and skywaves can all achieve greater ranges than line-of-sight propagation. The study of radio propagation allows estimates of useful range to be made.

Frequency Bands

Frequency	Wavelength	Designation	Abbreviation
3–30 Hz	10^5–10^4 km	Extremely low frequency	ELF
30–300 Hz	10^4–10^3 km	Super low frequency	SLF
300–3000 Hz	10^3–100 km	Ultra low frequency	ULF
3–30 kHz	100–10 km	Very low frequency	VLF
30–300 kHz	10–1 km	Low frequency	LF
300 kHz – 3 MHz	1 km – 100 m	Medium frequency	MF
3–30 MHz	100–10 m	High frequency	HF
30–300 MHz	10–1 m	Very high frequency	VHF
300 MHz – 3 GHz	1 m – 10 cm	Ultra high frequency	UHF
3–30 GHz	10–1 cm	Super high frequency	SHF
30–300 GHz	1 cm – 1 mm	Extremely high frequency	EHF
300 GHz – 3 THz	1 mm – 0.1 mm	Tremendously high frequency	THF

In Medicine

Radio frequency (RF) energy, in the form of radiating waves or electrical currents, has been used in medical treatments for over 75 years, generally for minimally invasive surgeries using radiofrequency ablation including the treatment of sleep apnea. Magnetic resonance imaging (MRI) uses radio frequency waves to generate images of the human body.

Radio frequencies at non-ablation energy levels are sometimes used as a form of cosmetic treatment that can tighten skin, reduce fat (lipolysis), or promote healing.

RF diathermy is a medical treatment that uses RF induced heat as a form of physical or occupational therapy and in surgical procedures. It is commonly used for muscle relaxation. It is also a method of heating tissue electromagnetically for therapeutic purposes in medicine. Diathermy is used in physical therapy and occupational therapy to deliver moderate heat directly to pathologic lesions in the deeper tissues of the body. Surgically, the extreme heat that can be produced by diathermy may be used to destroy neoplasms, warts, and infected tissues, and to cauterize blood vessels to prevent excessive bleeding. The technique is particularly valuable in neurosurgery and surgery of the eye. Diathermy equipment typically operates in the short-wave radio frequency (range 1–100 MHz) or microwave energy (range 434–915 MHz).

Pulsed electromagnetic field therapy (PEMF) is a medical treatment that purportedly helps to heal bone tissue reported in a recent NASA study. This method usually employs electromagnetic radiation of different frequencies - ranging from static magnetic fields, through extremely low frequencies (ELF) to higher radio frequencies (RF) administered in pulses.

Effects on the Human Body

Extremely Low Frequency RF

High-power extremely low frequency RF with electric field levels in the low kV/m range are known to induce perceivable currents within the human body that create an annoying tingling sensation. These currents will typically flow to ground through a body contact surface such as the feet, or arc to ground where the body is well insulated.

Microwaves

Microwave exposure at low-power levels below the Specific absorption rate set by government regulatory bodies are considered harmless non-ionizing radiation and have no effect on the human body. However, levels above the Specific absorption rate set by the U.S. Federal Communications Commission are considered potentially harmful.

Long-term human exposure to high-levels of microwaves is recognized to cause cataracts according to experimental animal studies and epidemiological studies. The mechanism is unclear but may include changes in heat sensitive enzymes that normally protect cell proteins in the lens. Another mechanism that has been advanced is direct damage to the lens from pressure waves induced in the aqueous humor.

High-power exposure to microwave RF is known to create a range of effects from lower to higher power levels, ranging from unpleasant burning sensation on the skin and microwave auditory effect, to extreme pain at the mid-range, to physical burning and blistering of skin and internals at high power levels.

General RF Exposure

The 1999 revision of Canadian Safety Code 6 recommended electric field limits of 100 kV/m for pulsed EMF to prevent air breakdown and spark discharges, mentioning rationale related to auditory effect and energy-induced unconsciousness in rats. The pulsed EMF limit was removed in later revisions, however.

As a Weapon

A heat ray is an RF harassment device that makes use of microwave radio frequencies to create an unpleasant heating effect in the upper layer of the skin. A publicly known heat ray weapon called the Active Denial System was developed by the US military as an experimental weapon to deny the enemy access to an area. A death ray is a weapon that delivers heat ray electromagnetic energy at levels that injure human tissue. The inventor of the death ray, Harry Grindell Matthews, claims to have lost sight in his left eye while developing his death ray weapon based on a primitive microwave magnetron from the 1920s (note that a typical microwave oven induces a tissue damaging cooking effect inside the oven at about 2 kV/m.)

Measurement

Since radio frequency radiation has both an electric and a magnetic component, it is often convenient to express intensity of radiation field in terms of units specific to each component. The unit volts per meter (V/m) is used for the electric component, and the unit amperes per meter (A/m) is used for the magnetic component. One can speak of an electromagnetic field, and these units are used to provide information about the levels of electric and magnetic field strength at a measurement location.

Another commonly used unit for characterizing an RF electromagnetic field is power density. Power density is most accurately used when the point of measurement is far enough away from the RF emitter to be located in what is referred to as the far field zone of the radiation pattern. In closer proximity to the transmitter, i.e., in the "near field" zone, the physical relationships between the electric and magnetic components of the field can be complex, and it is best to use the field strength units discussed above. Power density is measured in terms of power per unit area, for example, milliwatts per square centimeter (mW/cm²). When speaking of frequencies in the microwave range and higher, power density is usually used to express intensity since exposures that might occur would likely be in the far field zone.

References

- Hacking Exposed Linux: Linux Security Secrets & Solutions (third ed.). McGraw-Hill Osborne Media. 2008. p. 298. ISBN 978-0-07-226257-5.

- Sen, Dipankar; Sen, Prosenjit; Das, Anand M. (2009), RFID For Energy and Utility Industries, PennWell, ISBN 978-1-59370-105-5, pp. 1-48

- Katherine Albrecht; Liz McIntyre (2005). Spychips: how major corporations and government plan to track your every move with RFID. Thomas Nelson Inc. ISBN 1-59555-020-8.

- Ruey J. Sung & Michael R. Lauer (2000). Fundamental approaches to the management of cardiac arrhythmias. Springer. p. 153. ISBN 978-0-7923-6559-4.

- Melvin A. Shiffman, Sid J. Mirrafati, Samuel M. Lam and Chelso G. Cueteaux (2007). Simplified Facial Rejuvenation. Springer. p. 157. ISBN 978-3-540-71096-7.

- "Regulatory status for using RFID in the EPC Gen 2 band (860 to 960 MHz) of the UHF spectrum" (PDF). GS1.org. 2014-10-31. Retrieved 2015-03-23.

- "EPC™ Radio-Frequency Identity Protocols Generation-2 UHF RFID, Version 2.0.0" (PDF). GS1.org. November 2013. Retrieved 23 March 2015.

- "Mexico's Electronic Vehicle Registration system opens with Sirit open road toll technology, Dec 29, 2009". Tollroadsnews.com. Retrieved 2013-09-03.

- "RFID Technology Transforming Food Retailers Like Wal-Mart [Google Inc, Ingram Micro Inc.". Seeking Alpha. 2010-03-18. Retrieved 2013-09-22.

- Iain Thomson in San Francisco. "Hacker clones passports in drive-by RFID heist – V3.co.uk – formerly vnunet.com". V3.co.uk. Retrieved 2010-04-24.

- Rohrlich, Justin (15 December 2010). "RFID-Tagged Gaming Chips Render Hotel Bellagio Robbery Haul Worthless". Minyanville Financial Media. Retrieved 16 December 2010.

- O'Connor, Mary Catherine (March 18, 2009). Dash7 Alliance Seeks to Promote RFID Hardware Based on ISO 18000-7 Standard. RFID Journal LLC. Retrieved 2010-03-23.

2

Major Processes of Radio Frequency Identification

Radio frequency is an electromagnetic frequency that lies between 3 kHz to 300 GHz. In the process of radio frequency identification, the tracking label exerts an electromagnetic field that is emitted in the form of radio waves that communicate data. The whole process use near-field communication to carry out the transmission of information. The chapter studies electromagnetic field, radio waves and near-field communication minutely.

Electromagnetic Field

An electromagnetic field (also EM field) is a physical field produced by electrically charged objects. It affects the behavior of charged objects in the vicinity of the field. The electromagnetic field extends indefinitely throughout space and describes the electromagnetic interaction. It is one of the four fundamental forces of nature (the others are gravitation, weak interaction and strong interaction).

The field can be viewed as the combination of an electric field and a magnetic field. The electric field is produced by stationary charges, and the magnetic field by moving charges (currents); these two are often described as the sources of the field. The way in which charges and currents interact with the electromagnetic field is described by Maxwell's equations and the Lorentz force law.

From a classical perspective in the history of electromagnetism, the electromagnetic field can be regarded as a smooth, continuous field, propagated in a wavelike manner; whereas from the perspective of quantum field theory, the field is seen as quantized, being composed of individual particles.

Structure

The electromagnetic field may be viewed in two distinct ways: a continuous structure or a discrete structure.

Continuous Structure

Classically, electric and magnetic fields are thought of as being produced by smooth motions of charged objects. For example, oscillating charges produce electric and magnetic fields that may be viewed in a 'smooth', continuous, wavelike fashion. In this case, energy is viewed as being transferred continuously through the electromagnetic field between any two locations. For instance, the metal atoms in a radio transmitter appear to transfer energy continuously. This view is useful to a certain extent (radiation of low frequency), but problems are found at high frequencies.

Discrete Structure

The electromagnetic field may be thought of in a more 'coarse' way. Experiments reveal that in some circumstances electromagnetic energy transfer is better described as being carried in the form of packets called quanta (in this case, photons) with a fixed frequency. Planck's relation links the energy E of a photon to its frequency ν through the equation:

where h is Planck's constant, and ν is the frequency of the photon . Although modern quantum optics tells us that there also is a semi-classical explanation of the photoelectric effect—the emission of electrons from metallic surfaces subjected to electromagnetic radiation—the photon was historically (although not strictly necessarily) used to explain certain observations. It is found that increasing the intensity of the incident radiation (so long as one remains in the linear regime) increases only the number of electrons ejected, and has almost no effect on the energy distribution of their ejection. Only the frequency of the radiation is relevant to the energy of the ejected electrons.

This quantum picture of the electromagnetic field (which treats it as analogous to harmonic oscillators) has proved very successful, giving rise to quantum electrodynamics, a quantum field theory describing the interaction of electromagnetic radiation with charged matter. It also gives rise to quantum optics, which is different from quantum electrodynamics in that the matter itself is modelled using quantum mechanics rather than quantum field theory.

Dynamics

In the past, electrically charged objects were thought to produce two different, unrelated types of field associated with their charge property. An electric field is produced when the charge is stationary with respect to an observer measuring the properties of the charge, and a magnetic field as well as an electric field is produced when the charge moves, creating an electric current with respect to this observer. Over time, it was realized that the electric and magnetic fields are better thought of as two parts of a greater whole — the electromagnetic field. Until 1820, when the Danish physicist H. C. Ørsted discovered the effect of electricity through a wire on a compass needle, electricity and magnetism had been viewed as unrelated phenomena. In 1831, Michael Faraday, one of the great thinkers of his time, made the seminal observation that time-varying magnetic fields could induce electric currents and then, in 1864, James Clerk Maxwell published his famous paper A Dynamical Theory of the Electromagnetic Field.

Once this electromagnetic field has been produced from a given charge distribution, other charged objects in this field will experience a force in a similar way that planets experience a force in the gravitational field of the sun. If these other charges and currents are comparable in size to the sources producing the above electromagnetic field, then a new net electromagnetic field will be produced. Thus, the electromagnetic field may be viewed as a dynamic entity that causes other charges and currents to move, and which is also affected by them. These interactions are described by Maxwell's equations and the Lorentz force law. This discussion ignores the radiation reaction force.

Feedback Loop

The behavior of the electromagnetic field can be divided into four different parts of a loop:

- the electric and magnetic fields are generated by electric charges,
- the electric and magnetic fields interact with each other,
- the electric and magnetic fields produce forces on electric charges,
- the electric charges move in space.

A common misunderstanding is that (a) the quanta of the fields act in the same manner as (b) the charged particles that generate the fields. In our everyday world, charged particles, such as electrons, move slowly through matter with a drift velocity of a fraction of a centimeter (or inch) per second, but fields propagate at the speed of light - approximately 300 thousand kilometers (or 186 thousand miles) a second. The mundane speed difference between charged particles and field quanta is on the order of one to a million, more or less. Maxwell's equations relate (a) the presence and movement of charged particles with (b) the generation of fields. Those fields can then affect the force on, and can then move other slowly moving charged particles. Charged particles can move at relativistic speeds nearing field propagation speeds, but, as Einstein showed, this requires enormous field energies, which are not present in our everyday experiences with electricity, magnetism, matter, and time and space.

The feedback loop can be summarized in a list, including phenomena belonging to each part of the loop:

- charged particles generate electric and magnetic fields
- the fields interact with each other
  - changing electric field acts like a current, generating 'vortex' of magnetic field
  - Faraday induction: changing magnetic field induces (negative) vortex of electric field

- o   Lenz's law: negative feedback loop between electric and magnetic fields

- fields act upon particles

    - o   Lorentz force: force due to electromagnetic field

        - electric force: same direction as electric field

        - magnetic force: perpendicular both to magnetic field and to velocity of charge

- particles move

    - o   current is movement of particles

- particles generate more electric and magnetic fields; cycle repeats

Mathematical Description

There are different mathematical ways of representing the electromagnetic field. The first one views the electric and magnetic fields as three-dimensional vector fields. These vector fields each have a value defined at every point of space and time and are thus often regarded as functions of the space and time coordinates. As such, they are often written as E(x, y, z, t) (electric field) and B(x, y, z, t) (magnetic field).

If only the electric field (E) is non-zero, and is constant in time, the field is said to be an electrostatic field. Similarly, if only the magnetic field (B) is non-zero and is constant in time, the field is said to be a magnetostatic field. However, if either the electric or magnetic field has a time-dependence, then both fields must be considered together as a coupled electromagnetic field using Maxwell's equations.

With the advent of special relativity, physical laws became susceptible to the formalism of tensors. Maxwell's equations can be written in tensor form, generally viewed by physicists as a more elegant means of expressing physical laws.

The behaviour of electric and magnetic fields, whether in cases of electrostatics, magnetostatics, or electrodynamics (electromagnetic fields), is governed by Maxwell's equations. In the vector field formalism, these are:

$$\nabla \cdot \mathbf{E} = \frac{\rho}{\varepsilon_0} \text{ (Gauss's law)}$$

$$\nabla \cdot \mathbf{B} = 0 \text{ (Gauss's law for magnetism)}$$

$$\nabla \times \mathbf{E} = -\frac{\partial \mathbf{B}}{\partial t} \text{ (Faraday's law)}$$

$$\nabla \times \mathbf{B} = \mu_0 \mathbf{J} + \mu_0 \varepsilon_0 \frac{\partial \mathbf{E}}{\partial t} \text{ (Maxwell–Ampère law)}$$

where ρ is the charge density, which can (and often does) depend on time and position, ϵ_0 is the permittivity of free space, μ_0 is the permeability of free space, and J is the current density vector, also a function of time and position. The units used above are the standard SI units. Inside a linear material, Maxwell's equations change by switching the permeability and permittivity of free space with the permeability and permittivity of the linear material in question. Inside other materials which possess more complex responses to electromagnetic fields, these terms are often represented by complex numbers, or tensors.

The Lorentz force law governs the interaction of the electromagnetic field with charged matter.

When a field travels across to different media, the properties of the field change according to the various boundary conditions. These equations are derived from Maxwell's equations. The tangential components of the electric and magnetic fields as they relate on the boundary of two media are as follows:

$$\mathbf{E}_1 = \mathbf{E}_2$$

$$\mathbf{H}_1 = \mathbf{H}_2 \text{ (current-free)}$$

$$\mathbf{D}_1 = \mathbf{D}_2 \text{ (charge-free)}$$

$$\mathbf{B}_1 = \mathbf{B}_2$$

The angle of refraction of an electric field between media is related to the permittivity (ε) of each medium:

$$\frac{\tan \theta_1}{\tan \theta_2} \quad \frac{\varepsilon_2}{\varepsilon_1}$$

The angle of refraction of a magnetic field between media is related to the permeability of each medium:

$$\frac{\tan \theta_1}{\tan \theta_2} = \frac{\mu_{r2}}{\mu_{r1}}$$

Properties of the Field

Reciprocal Behavior of Electric and Magnetic Fields

The two Maxwell equations, Faraday's Law and the Ampère-Maxwell Law, illustrate a very practical feature of the electromagnetic field. Faraday's Law may be stated roughly as 'a changing magnetic field creates an electric field'. This is the principle behind the electric generator.

Ampere's Law roughly states that 'a changing electric field creates a magnetic field'. Thus, this law can be applied to generate a magnetic field and run an electric motor.

Light as an Electromagnetic Disturbance

Maxwell's equations take the form of an electromagnetic wave in a volume of space not containing charges or currents (free space) – that is, where ρ and J are zero. Under these conditions, the electric and magnetic fields satisfy the electromagnetic wave equation:

$$\left(\nabla^2 - \frac{1}{c^2}\frac{\partial^2}{\partial t^2}\right)\mathbf{E} = 0$$

$$\left(\nabla^2 - \frac{1}{c^2}\frac{\partial^2}{\partial t^2}\right)\mathbf{B} = 0$$

James Clerk Maxwell was the first to obtain this relationship by his completion of Maxwell's equations with the addition of a displacement current term to Ampere's Circuital law.

Relation to and Comparison with other Physical Fields

Being one of the four fundamental forces of nature, it is useful to compare the electromagnetic field with the gravitational, strong and weak fields. The word 'force' is sometimes replaced by 'interaction' because modern particle physics models electromagnetism as an exchange of particles known as gauge bosons.

Electromagnetic and Gravitational Fields

Sources of electromagnetic fields consist of two types of charge – positive and negative. This contrasts with the sources of the gravitational field, which are masses. Masses are sometimes described as gravitational charges, the important feature of them being that there are only positive masses and no negative masses. Further, gravity differs from electromagnetism in that positive masses attract other positive masses whereas same charges in electromagnetism repel each other.

The relative strengths and ranges of the four interactions and other information are tabulated below:

Theory	Interaction	mediator	Relative Magnitude	Behavior	Range
Chromodynamics	Strong interaction	gluon	10^{38}	1	10^{-15} m
Electrodynamics	Electromagnetic interaction	photon	10^{36}	$1/r^2$	infinite
Flavordynamics	Weak interaction	W and Z bosons	10^{25}	$1/r^5$ to $1/r^7$	10^{-16} m

Geometrody-namics	Gravitation	graviton	10^0	$1/r^2$	infinite

Applications

Static E and M Fields and Static EM Fields

When an EM field is not varying in time, it may be seen as a purely electrical field or a purely magnetic field, or a mixture of both. However the general case of a static EM field with both electric and magnetic components present, is the case that appears to most observers. Observers who see only an electric or magnetic field component of a static EM field, have the other (electric or magnetic) component suppressed, due to the special case of the immobile state of the charges that produce the EM field in that case. In such cases the other component becomes manifest in other observer frames.

A consequence of this, is that any case that seems to consist of a "pure" static electric or magnetic field, can be converted to an EM field, with both E and M components present, by simply moving the observer into a frame of reference which is moving with regard to the frame in which only the "pure" electric or magnetic field appears. That is, a pure static electric field will show the familiar magnetic field associated with a current, in any frame of reference where the charge moves. Likewise, any new motion of a charge in a region that seemed previously to contain only a magnetic field, will show that that the space now contains an electric field as well, which will be found to produces an additional Lorentz force upon the moving charge.

Thus, electrostatics, as well as magnetism and magnetostatics, are now seen as studies of the static EM field when a particular frame has been selected to suppress the other type of field, and since an EM field with both electric and magnetic will appear in any other frame, these "simpler" effects are merely the observer's. The "applications" of all such non-time varying (static) fields are discussed in the main articles linked in this section.

Time-varying EM Fields in Maxwell's Equations

An EM field that varies in time has two "causes" in Maxwell's equations. One is charges and currents (so-called "sources"), and the other cause for an E or M field is a change in the other type of field (this last cause also appears in "free space" very far from currents and charges).

An electromagnetic field very far from currents and charges (sources) is called electromagnetic radiation (EMR) since it radiates from the charges and currents in the source, and has no "feedback" effect on them, and is also not affected directly by them in the present time (rather, it is indirectly produced by a sequences of changes in fields radiating out from them in the past). EMR consists of the radiations in the electromagnetic spectrum, including radio waves, microwave, infrared, visible light, ultraviolet light, X-rays, and gamma rays. The many commercial applications of these radiations are discussed in the named and linked articles.

A notable application of visible light is that this type of energy from the Sun powers all life on Earth that either makes or uses oxygen.

A changing electromagnetic field which is physically close to currents and charges will have a dipole characteristicthat is dominated by either a changing electric dipole, or a changing magnetic dipole. This type of dipole field near sources is called an electro-magnetic near-field.

Changing electric dipole fields, as such, are used commercially as near-fields mainly as a source of dielectric heating. Otherwise, they appear parasitically around conductors which absorb EMR, and around antennas which have the purpose of generating EMR at greater distances.

Changing magnetic dipole fields (i.e., magnetic near-fields) are used commercially for many types of magnetic induction devices. These include motors and electrical trans-formers at low frequencies, and devices such as metal detectors and MRI scanner coils at higher frequencies. Sometimes these high-frequency magnetic fields change at radio frequencies without being far-field waves and thus radio waves. Further uses of near-field EM effects commercially, may be found in the article on virtual photons, since at the quantum level, these fields are rep-resented by these particles. Far-field effects (EMR) in the quantum picture of radiation, are represented by ordinary photons.

Health and Safety

The potential health effects of the very low frequency EMFs surrounding power lines and electrical devices are the subject of on-going research and a significant amount of public debate. The US National Institute for Occupational Safety and Health (NIOSH) and other US government agencies do not consider EMFs a proven health hazard. NIOSH has issued some cautionary advisories but stresses that the data are currently too limited to draw good conclusions.

The potential effects of electromagnetic fields on human health vary widely de-pending on the frequency and intensity of the fields. For more information on the health effects due to specific parts of the electromagnetic spectrum, see the follow-ing articles:

- Static electric fields

- Static magnetic fields

- Extremely low frequency (ELF)

- Radio frequency (RF)

- Light

- Ultraviolet (UV)

- Gamma rays

- Mobile telephony

Radio Wave

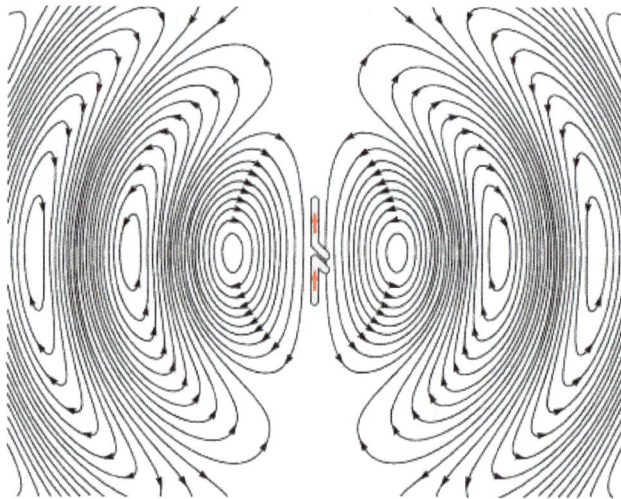

Animation of a half-wave dipole antenna radiating radio waves, showing the electric field lines. The antenna in the center is two vertical metal rods, with an alternating current applied at its center from a radio transmitter (not shown). The voltage charges the two sides of the antenna alternately positive (+) and negative (−). Loops of electric field (black lines) leave the antenna and travel away at the speed of light; these are the radio waves. The action is drastically slowed down in this animation.

Radio waves are a type of electromagnetic radiation with wavelengths in the electromagnetic spectrum longer than infrared light. Radio waves have frequencies as high as 300 GHz to as low as 3 kHz, though some definitions describe waves above 1 or 3 GHz as microwaves, or include waves of any lower frequency. At 300 GHz, the corresponding wavelength is 1 mm (0.039 in), and at 3 kHz is 100 km (62 mi). Like all other electromagnetic waves, they travel at the speed of light. Naturally occurring radio waves are made by lightning, or by astronomical objects. Artificially generated radio waves are used for fixed and mobile radio communication, broadcasting, radar and other navigation systems, communications satellites, computer networks and innumerable other applications. Radio waves are generated by radio transmitters and received by radio receivers. Different frequencies of radio waves have different propagation characteristics in the Earth's atmosphere; long waves can diffract around obstacles like mountains and follow the contour of the earth (ground waves), shorter waves can reflect off the ionosphere and return to earth beyond the horizon (skywaves), while much shorter wavelengths bend or diffract very little and travel on a line of sight, so their propagation distances are limited to the visual horizon.

To prevent interference between different users, the artificial generation and use of radio waves is strictly regulated by law, coordinated by an international body called the International Telecommunications Union (ITU), which defines radio waves as "electromagnetic waves of frequencies arbitrarily lower than 3 000 GHz, propagated in space without artificial guide". The radio spectrum is divided into a number of radio bands on the basis of frequency, allocated to different uses.

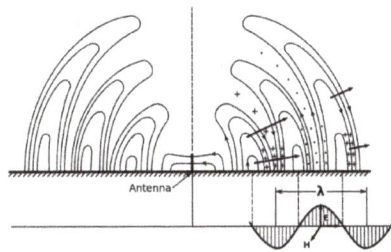

Diagram of the electric fields (E) and magnetic fields (H) of radio waves emitted by a monopole radio transmitting antenna (small dark vertical line in the center). The E and H fields are perpendicular as implied by the phase diagram in the lower right.

Discovery and Utilization

Radio waves were first predicted by mathematical work done in 1867 by Scottish mathematical physicist James Clerk Maxwell. Maxwell noticed wavelike properties of light and similarities in electrical and magnetic observations. His mathematical theory, now called Maxwell's equations, described light waves and radio waves as waves of electromagnetism that travel in space, radiated by a charged particle as it undergoes acceleration. In 1887, Heinrich Hertz demonstrated the reality of Maxwell's electromagnetic waves by experimentally generating radio waves in his laboratory, showing that they exhibited the same wave properties as light: standing waves, refraction, diffraction, and polarization. Radio waves were first used for communication in the mid 1890s by Guglielmo Marconi, who developed the first practical radio transmitters and receivers.

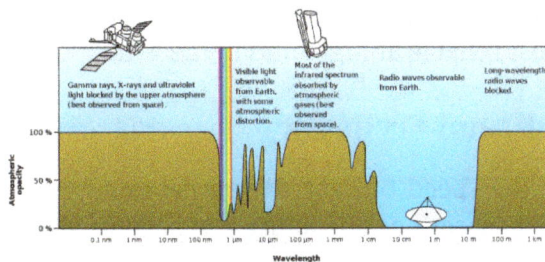

Rough plot of Earth's atmospheric transmittance (or opacity) to various wavelengths of electromagnetic radiation, including radio waves.

Propagation

The study of electromagnetic phenomena such as reflection, refraction, polarization, diffraction, and absorption is of critical importance in the study of how radio waves move in free space and over the surface of the Earth. Different frequencies experience different combinations of these phenomena in the Earth's atmosphere, making certain radio bands more useful for specific purposes than others. Practical radio systems use three different techniques of radio propagation to communicate:

- Line of sight: This refers to radio waves that travel in a straight line from the transmitting antenna to the receiving antenna. It does not necessarily require a cleared sight path; at lower frequencies radio waves can pass though buildings and other obstructions. This is the only method of propagation possible at microwave frequencies. On the surface of the Earth, line of sight propagation is limited by the visual horizon to about 40 miles. It is the method used in short range radio communication systems such as cell phones, walkie-talkies, wireless networks, FM and television broadcasting and radar. By using dish antennas to transmit beams of microwaves, point-to-point radio relay links transmit telephone and television signals over long distances up to the visual horizon, and ground stations can communicate with satellites and spacecraft billions of miles from Earth.

- Ground waves: At lower frequencies, in the medium wave and longwave bands, due to diffraction radio waves can bend over hills and mountains, and beyond the horizon, traveling as surface waves which follow the contour of the Earth. This allows mediumwave and longwave broadcasting stations to have coverage areas of hundreds of miles. As the frequency drops, the losses decrease and the achievable range increases. Military very low frequency (VLF) communication systems can communicate over most of the Earth, and with submarines hundreds of feet underwater.

- Skywaves: At medium wave and shortwave wavelengths, radio waves reflect off a conductive ionized layer in the atmosphere called the ionosphere. So radio waves directed at an angle into the sky can return to Earth beyond the horizon; this is called "skip" or "skywave" propagation. By using multiple skips communication at intercontinental distances can be achieved. This technique is used by radio amateurs, and shortwave broadcasting stations to broadcast to other countries. Skywave propagation is variable and dependent on atmospheric conditions; it is most reliable at night and in the winter.

Speed, Wavelength, and Frequency

Radio waves travel at the speed of light. When passing through an object, they are slowed according to that object's permeability and permittivity.

The wavelength is the distance from one peak of the wave's electric field (wave's peak/crest) to the next, and is inversely proportional to the frequency of the wave. The distance a radio wave travels in one second, in a vacuum, is 299,792,458 meters (983,571,056 ft) which is the wavelength of a 1 hertz radio signal. A 1 megahertz radio signal has a wavelength of 299.8 meters (984 ft).

Radio Communication

In order to receive radio signals, for instance from AM/FM radio stations, a radio antenna must be used. However, since the antenna will pick up thousands of radio signals at a time, a radio tuner is necessary to tune in a particular signal. This is typically done via a resonator (in its simplest form, a circuit with a capacitor, inductor, or crystal oscillator, but many modern radios use Phase Locked Loop systems). The resonator is configured to resonate at a particular frequency, allowing the tuner to amplify sine waves at that radio frequency and ignore other sine waves. Usually, either the inductor or the capacitor of the resonator is adjustable, allowing the user to change the frequency at which it resonates.

Near Field Communication

Near-field communication (NFC) is a set of communication protocols that enable two electronic devices, one of which is usually a portable device such as a smartphone, to establish communication by bringing them within about 4 cm (2 in) of each other.

Overviewt

Similar ideas in advertising and industrial applications were not generally successful commercially, outpaced by technologies such as barcodes and UHF RFID tags. NFC protocols established a generally-supported standard. When one of the connected devices has Internet connectivity, the other can exchange data with online services.

NFC-enabled portable devices can be provided with apps, for example to read electronic tags or make payments when connected to an NFC-compliant apparatus. Earlier close-range communication used technology that was proprietary to the manufacturer, for applications such as stock ticket, access control and payment readers.

Like other "proximity card" technologies, NFC employs electromagnetic induction between two loop antennae when NFC devices—for example a smartphone and a "smartposter"—exchange information, operating within the globally available unlicensed radio frequency ISM band of 13.56 MHz on ISO/IEC 18000-3 air interface at rates ranging from 106 to 424 kbit/s.

Each full NFC device can work in three modes:

- NFC card emulation—enables NFC-enabled devices such as smartphones to act like smart cards, allowing users to perform transactions such as payment or ticketing.

- NFC reader/writer—enables NFC-enabled devices to read information stored on inexpensive NFC tags embedded in labels or smart posters.

- NFC peer-to-peer—enables two NFC-enabled devices to communicate with each other to exchange information in an adhoc fashion.

NFC tags are passive data stores which can be read, and under some circumstances written to, by an NFC device. They typically contain data (as of 2015 between 96 and 8,192 bytes) and are read-only in normal use, but may be rewritable. Applications include secure personal data storage (e.g. debit or credit card information, loyalty program data, personal identification numbers (PINs), contacts). NFC tags can be custom-encoded by their manufacturers or use the industry specifications.

The standards were provided by the NFC Forum. The forum was responsible for promoting the technology and setting standards and certifies device compliance. Secure communications are available by applying encryption algorithms as is done for Credit Card and if it fits the criteria for being considered a personal area network.

NFC standards cover communications protocols and data exchange formats and are based on existing radio-frequency identification (RFID) standards including ISO/IEC 14443 and FeliCa. The standards include ISO/IEC 18092 and those defined by the NFC Forum. In addition to the NFC Forum, the GSMA group defined a platform for the deployment of within mobile handsets. GSMA's efforts include Trusted Services Manager, Single Wire Protocol, testing/certification and secure element.

A patent licensing program for NFC is under deployment by France Brevets, a patent fund created in 2011. This program was under development by Via Licensing Corporation, an independent subsidiary of Dolby Laboratories, and was terminated in May 2012. A platform-independent free and open source NFC library, libnfc, is available under the GNU Lesser General Public License.

Present and anticipated applications include contactless transactions, data exchange and simplified setup of more complex communications such as Wi-Fi.

History

NFC is rooted in radio-frequency identification technology (known as RFID) which allows compatible hardware to both supply power to and communicate with an otherwise unpowered and passive electronic tag using radio waves. This is used for identification, authentication and tracking.

- 1983 The first patent to be associated with the abbreviation "RFID" was granted to Charles Walton.

- 1997 Early form patented and first used in Star Wars character toys for Hasbro. The patent was originally held by Andrew White and Marc Borrett at Innovision Research and Technology (Patent WO9723060). The device allowed data communication between two units in close proximity.

- 2002 Sony and Philips agreed to establish a technology specification and created a technical outline on March 25, 2002.

- 2003 NFC was approved as an ISO/IEC standard on December 8, and later as an ECMA standard.

- 2004 Nokia, Philips and Sony established the NFC Forum

- 2006 Initial specifications for NFC Tags

- 2006 Specification for "SmartPoster" records

- 2007 Innovision's NFC tags used in the first consumer trial in the UK, in the Nokia 6131 handset.

- 2009 In January, NFC Forum released Peer-to-Peer standards to transfer contacts, URLs, initiate Bluetooth, etc.

- 2010 Innovision released a suite of designs and patents for low cost, mass-market mobile phones and other devices.

- 2010 Samsung Nexus S: First Android NFC phone shown

- 2010 Nice, France launches the "Nice City of contactless mobile" project, providing inhabitants with NFC mobile phones and bank cards, and a "bouquet of services" covering transportation, tourism and student's services

- 2011 Tapit Media launches in Sydney, Australia as the first specialized NFC marketing company

- 2011 Google I/O "How to NFC" demonstrates NFC to initiate a game and to share a contact, URL, app or video.

- 2011 NFC support becomes part of the Symbian mobile operating system with the release of Symbian Anna version.

- 2011 Research In Motion devices are the first ones certified by MasterCard Worldwide for their PayPass service

- 2012 UK restaurant chain EAT. and Everything Everywhere (Orange Mobile Network Operator), partner on the UK's first nationwide NFC-enabled smartposter campaign. A specially created mobile phone app is triggered when the NFC-enabled mobile phone comes into contact with the smartposter.

- 2012 Sony introduced NFC "Smart Tags" to change modes and profiles on a Sony smartphone at close range, included with the Sony Xperia P Smartphone released the same year.

- 2013 Samsung and VISA announce their partnership to develop mobile payments.

- 2013 IBM scientists, in an effort to curb fraud and security breaches, develop an NFC-based mobile authentication security technology. This technology works on similar principles to dual-factor authentication security.

- 2014 AT&T, Verizon and T-Mobile released Softcard (formally ISIS mobile wallet). It runs on NFC-enabled Android phones and iPhone 4 and iPhone 5 when an external NFC case is attached. The technology was purchased by Google and the service ended on March 31, 2015.

- 2014 Apple introduced Apple Pay for NFC-enabled mobile payment on iPhone 6 and 6 Plus, and the Apple Watch, which was released on April 24, 2015.

- In November 2015, Swatch and Visa Inc. announced a partnership to enable NFC financial transactions using the "Swatch Bellamy" wristwatch. The system is currently online in Asia thanks to a partnership with China UnionPay and Bank of Communications. The partnership will bring the technology to the US, Brazil, and Switzerland.

Design

NFC is a set of short-range wireless technologies, typically requiring a separation of 10 cm or less. NFC operates at 13.56 MHz on ISO/IEC 18000-3 air interface and at rates ranging from 106 kbit/s to 424 kbit/s. NFC always involves an initiator and a target; the initiator actively generates an RF field that can power a passive target. This enables NFC targets to take very simple form factors such as unpowered tags, stickers, key fobs, or cards. NFC peer-to-peer communication is possible, provided both devices are powered.

NFC tags contain data and are typically read-only, but may be writeable. They can be custom-encoded by their manufacturers or use NFC Forum specifications. The tags can securely store personal data such as debit and credit card information, loyalty program data, PINs and networking contacts, among other information. The NFC Forum defines four types of tags that provide different communication speeds and capabilities in terms of configurability, memory, security, data retention and write endurance. Tags currently offer between 96 and 4,096 bytes of memory.

As with proximity card technology, near-field communication uses magnetic induction between two loop antennas located within each other's near field, effectively forming an air-core transformer. It operates within the globally available and unlicensed radio frequency ISM band of 13.56 MHz. Most of the RF energy is concentrated in the allowed ±7 kHz bandwidth range, but the full spectral envelope may be as wide as 1.8 MHz when using ASK modulation.

Theoretical working distance with compact standard antennas: up to 20 cm (practical working distance of about 10 cm).

Supported data rates: 106, 212 or 424 kbit/s (the bit rate 848 kbit/s is not compliant with the standard ISO/IEC 18092)

The two modes are:

- Passive—The initiator device provides a carrier field and the target device answers by modulating the existing field. In this mode, the target device may draw its operating power from the initiator-provided electromagnetic field, thus making the target device a transponder.

- Active—Both initiator and target device communicate by alternately generating their own fields. A device deactivates its RF field while it is waiting for data. In this mode, both devices typically have power supplies.

Speed	Active device	Passive device
424 kbit/s	Man, 10% ASK	Man, 10% ASK
212 kbit/s	Man, 10% ASK	Man, 10% ASK
106 kbit/s	Modified Miller, 100% ASK	Man, 10% ASK

NFC employs two different codings to transfer data. If an active device transfers data at 106 kbit/s, a modified Miller coding with 100% modulation is used. In all other cases Manchester coding is used with a modulation ratio of 10%.

NFC devices are full-duplex—they are able to receive and transmit data at the same time. Thus, they can check for potential collisions if the received signal frequency does not match the transmitted signal's frequency.

Although the range of NFC is limited to a few centimeters, plain NFC does not ensure secure communications. In 2006, Ernst Haselsteiner and Klemens Breitfuß described possible attacks and detailed how to leverage NFC's resistance to man-in-the-middle attacks to establish a specific key. As this technique is not part of the ISO standard, NFC offers no protection against eavesdropping and can be vulnerable to data modifications. Applications may use higher-layer cryptographic protocols (e.g. SSL) to establish a secure channel.

The RF signal for the wireless data transfer can be picked up with antennas. The distance from which an attacker is able to eavesdrop the RF signal depends on multiple parameters, but is typically less than 10 meters. Also, eavesdropping is highly affected by the communication mode. A passive device that doesn't generate its own RF field is much harder to eavesdrop on than an active device. An attacker can typically eavesdrop within 10 m and 1 m for active devices and passive devices, respectively.

Because NFC devices usually include ISO/IEC 14443 protocols, relay attacks are feasible.For this attack the adversary forwards the request of the reader to the victim and relays its answer to the reader in real time, pretending to be the owner of the victim's smart card. This is similar to a man-in-the-middle attack. One libnfc code example demonstrates a relay attack using two stock commercial NFC devices. This attack can be implemented using only two NFC-enabled mobile phones.

Standards

NFC standards cover communications protocols and data exchange formats, and are based on existing RFID standards including ISO/IEC 14443 and FeliCa. The standards include ISO/IEC 18092 and those defined by the NFC Forum.

NFC Protocol stack overview

ISO/IEC

NFC is standardized in ECMA-340 and ISO/IEC 18092. These standards specify the modulation schemes, coding, transfer speeds and frame format of the RF interface of NFC devices, as well as initialization schemes and conditions required for data collision-control during initialization for both passive and active NFC modes. They also define the transport protocol, including protocol activation and data-exchange methods. The air interface for NFC is standardized in:

- ISO/IEC 18092/ECMA-340—Near Field Communication Interface and Protocol-1 (NFCIP-1)

- ISO/IEC 21481/ECMA-352—Near Field Communication Interface and Protocol-2 (NFCIP-2)

NFC incorporates a variety of existing standards including ISO/IEC 14443 Type A and Type B, and FeliCa. NFC-enabled phones work at a basic level with existing readers. In "card emulation mode" an NFC device should transmit, at a minimum, a unique

ID number to a reader. In addition, NFC Forum defined a common data format called NFC Data Exchange Format (NDEF) that can store and transport items ranging from any MIME-typed object to ultra-short RTD-documents, such as URLs. The NFC Forum added the Simple NDEF Exchange Protocol (SNEP) to the spec that allows sending and receiving messages between two NFC devices.

Gsma

The GSM Association (GSMA) is a trade association representing nearly 800 mobile telephony operators and more than 200 product and service companies across 219 countries. Many of its members have led NFC trials and are preparing services for commercial launch.

GSM is involved with several initiatives:

- Standards: GSMA is developing certification and testing standards to ensure global interoperability of NFC services.

- Pay-Buy-Mobile initiative: Seeks to define a common global approach to using NFC technology to link mobile devices with payment and contactless systems.

- On November 17, 2010, after two years of discussions, AT&T, Verizon and T-Mobile launched a joint venture to develop a platform through which point of sale payments could be made using NFC in cell phones. Initially known as Isis Mobile Wallet and later as Softcard, the venture was designed to usher in broad deployment of NFC technology, allowing their customers' NFC-enabled cell phones to function similarly to credit cards throughout the US. Following an agreement with—and IP purchase by—Google, the Softcard payment system was shuttered in March, 2015, with an endorsement for its earlier rival, Google Wallet.

Stolpan

StoLPaN ('Store Logistics and Payment with NFC) is a pan-European consortium supported by the European Commission's Information Society Technologies program. StoLPaN will examine the potential for NFC local wireless mobile communication.

Nfc Forum

NFC Forum is a non-profit industry association formed on March 18, 2004, by NXP Semiconductors, Sony and Nokia to advance the use of NFC wireless interaction in consumer electronics, mobile devices and PCs. Standards include the four distinct tag types that provide different communication speeds and capabilities covering flexibility, memory, security, data retention and write endurance. NFC Forum promotes implementation and standardization of NFC technology to ensure interoperability between devices and services. As of June 2013, the NFC Forum had over 190 member companies.

NFC Forum promotes NFC and certifies device compliance and whether it fits in a personal area network.

Other Standardization Bodies

GSMA defined a platform for the deployment of GSMA NFC Standards within mobile handsets. GSMA's efforts include, Single Wire Protocol, testing and certification and secure element. The GSMA standards surrounding the deployment of NFC protocols (governed by NFC Forum) on mobile handsets are neither exclusive nor universally accepted. For example, Google's deployment of Host Card Emulation on Android KitKat provides for software control of a universal radio. In this HCE Deployment the NFC protocol is leveraged without the GSMA standards.

Other standardization bodies involved in NFC include:

- ETSI/SCP (Smart Card Platform) to specify the interface between the SIM card and the NFC chipset.

- GlobalPlatform to specify a multi-application architecture of the secure element.

- EMVCo for the impacts on the EMV payment applications

Applications

NFC allows one- and two-way communication between endpoints, suitable for many applications.

N-Mark logo for NFC-enabled devices

Commerce

NFC devices can be used in contactless payment systems, similar to those used in credit cards and electronic ticket smartcards and allow mobile payment to replace/supplement these systems.

In Android 4.4, Google introduced platform support for secure NFC-based transactions through Host Card Emulation (HCE), for payments, loyalty programs, card access, transit passes and other custom services. HCE allows any Android 4.4 app to emulate an NFC smart card, letting users initiate transactions with their device. Apps can use a new Reader Mode to act as readers for HCE cards and other NFC-based transactions.

On September 9, 2014, Apple announced support for NFC-powered transactions as part of Apple Pay. Apple stated that their approach to NFC payment is more secure because Apple Pay tokenizes its data to encrypt and protect it from unauthorized use.

Bootstrapping Other Connections

NFC offers a low-speed connection with simple setup that can be used to bootstrap more capable wireless connections. For example, Android Beam software uses NFC to enable pairing and establish a Bluetooth connection when doing a file transfer and then disabling Bluetooth on both devices upon completion. Nokia, Samsung, BlackBerry and Sony have used NFC technology to pair Bluetooth headsets, media players and speakers with one tap. The same principle can be applied to the configuration of Wi-Fi networks. Samsung Galaxy devices have a feature named S-Beam—an extension of Android Beam that uses NFC (to share MAC Address and IP addresses) and then uses Wi-Fi Direct to share files and documents. The advantage of using Wi-Fi Direct over Bluetooth is that it permits much faster data transfers, running up to 300Mbit/s.

Social Networking

NFC can be used for social networking, for sharing contacts, photos, videos or files and entering multiplayer mobile games.

Identity and Access Tokens

NFC-enabled devices can act as electronic identity documents and keycards. NFC's short range and encryption support make it more suitable than less private RFID systems.

Smartphone Automation and Nfc Tags

NFC-equipped smartphones can be paired with NFC Tags or stickers that can be programmed by NFC apps. These programs can allow a change of phone settings, texting, app launching, or command execution.

Such apps do not rely on a company or manufacturer, but can be utilized immediately with an NFC-equipped smartphone and an NFC tag.

The NFC Forum published the Signature Record Type Definition (RTD) 2.0 in 2015 to add integrity and authenticity for NFC Tags. This specification allows an NFC device to verify tag data and identify the tag author.

Gaming

NFC was used in video games starting with Skylanders: Spyro's Adventure. With it you buy figurines that are customizable and contain personal data with each figure, so no two figures are exactly alike. The Wii U was the first system to include NFC technology out of the box via the GamePad. It was later included in the New Nintendo 3DS range. The Amiibo range of accessories utilises NFC technology to unlock features.

Bluetooth Comparison

Aspect	NFC	Bluetooth	Bluetooth Low Energy
Tag requires power	No	Yes	Yes
Cost of Tag	$0.10 USD	$5.00 USD	$5.00 USD
RFID compatible	ISO 18000-3	Active	Active
Standardisation body	ISO/IEC	Bluetooth SIG	Bluetooth SIG
Network standard	ISO 13157 etc.	IEEE 802.15.1 (no longer maintained)	IEEE 802.15.1 (no longer maintained)
Network type	Point-to-point	WPAN	WPAN
Cryptography	Not with RFID	Available	Available
Range	< 20 cm	~100 m (class 1)	~50 m
Frequency	13.56 MHz	2.4–2.5 GHz	2.4–2.5 GHz
Bit rate	424 kbit/s	2.1 Mbit/s	1 Mbit/s
Set-up time	< 0.1 s	< 6 s	< 0.006 s
Current consumption	< 15mA (read)	Varies with class	< 15 mA (read and transmit)

NFC and Bluetooth are both short-range communication technologies available on mobile phones. NFC operates at slower speeds than Bluetooth, but consumes far less power and doesn't require pairing.

NFC sets up more quickly than standard Bluetooth, but has a lower transfer rate than Bluetooth low energy. With NFC, instead of performing manual configurations to identify devices, the connection between two NFC devices is automatically established in less than .1 second. The maximum data transfer rate of NFC (424 kbit/s) is slower than that of Bluetooth V2.1 (2.1 Mbit/s).

With a maximum working distance of less than 20 cm, NFC has a shorter range, which reduces the likelihood of unwanted interception. That makes NFC particularly suitable for crowded areas that complicate correlating a signal with its transmitting physical device (and by extension, its user).

NFC is compatible with existing passive RFID (13.56 MHz ISO/IEC 18000-3) infrastructures. NFC requires comparatively low power, similar to the Bluetooth V4.0 low energy protocol. When NFC works with an unpowered device (e.g. on a phone that may be turned off, a contactless smart credit card, a smart poster), however, the NFC power consumption is greater than that of Bluetooth V4.0 Low Energy, since illuminating the passive tag needs extra power.

Devices

In 2011, handset vendors released more than 40 NFC-enabled handsets with the Android mobile operating system.The iPhone 6 line is the first set of handsets from Apple to support NFC. BlackBerry devices support NFC using BlackBerry Tag on devices running BlackBerry OS 7.0 and greater.

Mastercard added further NFC support for PayPass for the Android and BlackBerry platforms, enabling PayPass users to make payments using their Android or BlackBerry smartphones. A partnership between Samsung and Visa added a 'payWave' application on the Galaxy S4 smartphone.

Microsoft added native NFC functionality in their mobile OS with Windows Phone 8, as well as the Windows 8 operating system. Microsoft provides the "Wallet hub" in Windows Phone 8 for NFC payment, and can integrate multiple NFC payment services within a single application.

Deployments

As of April 2011, hundreds of NFC trials had been conducted. Some firms moved to full-scale service deployments, spanning one or more countries. Multi-country deployments include Orange's rollout of NFC technology to banks, retailers, transport, and service providers in multiple European countries, and Airtel Africa and Oberthur Technologies deploying to 15 countries throughout Africa.

- China telecom (China's 3rd largest mobile operator) made its NFC rollout in November 2013. The company signed up multiple banks to make their payment apps available on its SIM Cards. China telecom stated that the wallet would support coupons, membership cards, fuel cards and boarding passes. The company planned to achieve targets of rolling out 40 NFC phone models and 30 Mn NFC SIMs by 2014.

- Softcard (formerly Isis Mobile Wallet), a joint venture from Verizon Wireless, AT&T and T-Mobile, focuses on in-store payments making use of NFC technology. After doing pilots in some regions, they launched across the US.

- Vodafone launched the NFC-based Vodafone SmartPass mobile payment service in Spain in partnership with Visa. It enables consumers with an NFC-enabled mobile device to make contactless payments via their SmartPass credit balance at any POS.

- OTI, an Israeli company that designs and develops contactless microprocessor based smart card technology, contracted to supply NFC-readers to one of its channel partners in the US. The partner was required to buy $10MM worth of OTI NFC readers over 3 years.

- Rogers Communications launched virtual wallet Suretap to enable users to make payments with their phone in Canada in April 2014. Suretap users can load up gift cards and pre-paid MasterCards from national retailers.

- Sri Lanka's first workforce smartcard, uses NFC.

- As of December 13, 2013 Tim Hortons TimmyME BlackBerry 10 Application allowed users to link their prepaid Tim Card to the app, allowing payment by tapping the NFC-enabled device to a standard contactless terminal.

- Google Wallet allows consumers to store credit card and store loyalty card information in a virtual wallet and then use an NFC-enabled device at terminals that also accept MasterCard PayPass transactions.

- Germany, Austria, Finland, New Zealand, Italy, Iran, and Turkey trialed NFC ticketing systems for public transport. The Lithuanian capital of Vilnius fully replaced paper tickets for public transportation with ISO/IEC 14443 Type A cards on July 1, 2013.

- NFC sticker-based payments in Australia's Bankmecu and card issuer Cuscal completed an NFC payment sticker trial, enabling consumers to make contactless payments at Visa payWave terminals using a smart sticker stuck to their phone.

- India was implementing NFC based transactions in box offices for ticketing purposes.

- A partnership of Google and Equity Bank in Kenya introduced NFC payment systems for public transport in the Capital city Nairobi under the branding "Beba Pay".

References

- Richard Feynman (1970). The Feynman Lectures on Physics Vol II. Addison Wesley Longman. ISBN 978-0-201-02115-8.

- Spencer, James N.; et al. (2010). Chemistry: Structure and Dynamics. John Wiley & Sons. p. 78. ISBN 9780470587119.

- Schaum's outline of theory and problems of electromagnetics(2nd Edition), Joseph A. Edminister, McGraw-Hill, 1995. ISBN 0070212341.

- Field and Wave Electromagnetics (2nd Edition), David K. Cheng, Prentice Hall, 1989. ISBN 978-0-201-12819-2.

- "NIOSH Fact Sheet: EMFs in the Workplace". United States National Institute for Occupational Safety and Health. 1996. Retrieved 31 August 2015.

3

Major Components of Radio Frequency Identification

The main components of radio frequency identification discussed in the chapter are integrated circuit, modulation and demodulation. The integrated circuit forms the transmitter tag that relays modulated signals which are then demodulated to bring about the original information. The chapter explores the classification and advances in integrated circuits, modulation methods and demodulation techniques.

Integrated Circuit

An integrated circuit or monolithic integrated circuit (also referred to as an IC, a chip, or a microchip) is a set of electronic circuits on one small plate ("chip") of semiconductor material, normally silicon. This can be made much smaller than a discrete circuit made from independent electronic components. ICs can be made very compact, having up to several billion transistors and other electronic components in an area the size of a human fingernail. The half-pitch between nodes in a circuit has been made smaller as the technology advances; in 2008 it dropped below 100 nanometers, and was reduced to around 14 nanometers in 2014.

Erasable programmable read-only memory integrated circuits. These packages have a transparent window that shows the die inside. The window allows the memory to be erased by exposing the chip to ultraviolet light.

ICs were made possible by experimental discoveries showing that semiconductor devices could perform the functions of vacuum tubes and by mid-20th-century technology advancements in semiconductor device fabrication. The integration of large numbers of tiny transistors into a small chip was an enormous improvement over the manual assembly of circuits using discrete electronic components. The integrated circuit's mass production capability, reliability and building-block approach to circuit design ensured the rapid adoption of standardized integrated circuits in place of designs using discrete transistors.

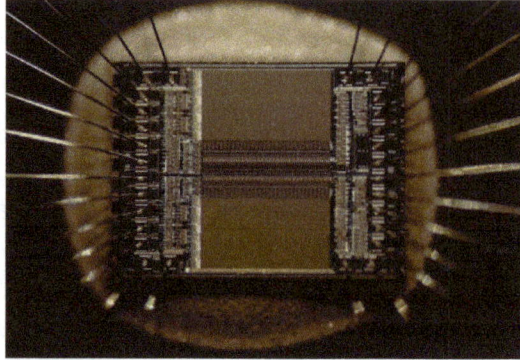

Integrated circuit from an EPROM memory microchip showing the memory blocks, the supporting circuitry and the fine silver wires which connect the integrated circuit die to the legs of the packaging.

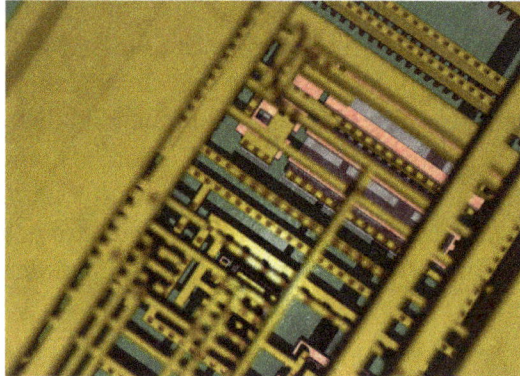

Synthetic detail of an integrated circuit through four layers of planarized copper interconnect, down to the polysilicon (pink), wells (greyish), and substrate (green)

ICs have two main advantages over discrete circuits: cost and performance. Cost is low because the chips, with all their components, are printed as a unit by photolithography rather than being constructed one transistor at a time. Furthermore, packaged ICs use much less material than discrete circuits. Performance is high because the IC's components switch quickly and consume little power (compared to their discrete counterparts) as a result of the small size and close proximity of the components. As of 2012, typical chip areas range from a few square millimeters to around 450 mm^2, with up to 9 million transistors per mm^2.

Integrated circuits are used in virtually all electronic equipment today and have revolutionized the world of electronics. Computers, mobile phones, and other digital home appliances are now inextricable parts of the structure of modern societies, made possible by the low cost of ICs.

Terminology

An integrated circuit is defined as:

A circuit in which all or some of the circuit elements are inseparably associated and electrically interconnected so that it is considered to be indivisible for the purposes of construction and commerce.

Circuits meeting this definition can be constructed using many different technologies, including thin-film transistor, thick film technology, or hybrid integrated circuit. However, in general usage integrated circuit has come to refer to the single-piece circuit construction originally known as a monolithic integrated circuit.

Invention

Early developments of the integrated circuit go back to 1949, when German engineer Werner Jacobi (Siemens AG) filed a patent for an integrated-circuit-like semiconductor amplifying device showing five transistors on a common substrate in a 3-stage amplifier arrangement. Jacobi disclosed small and cheap hearing aids as typical industrial applications of his patent. An immediate commercial use of his patent has not been reported.

The idea of the integrated circuit was conceived by Geoffrey W.A. Dummer (1909–2002), a radar scientist working for the Royal Radar Establishment of the British Ministry of Defence. Dummer presented the idea to the public at the Symposium on Progress in Quality Electronic Components in Washington, D.C. on 7 May 1952. He gave many symposia publicly to propagate his ideas, and unsuccessfully attempted to build such a circuit in 1956.

A precursor idea to the IC was to create small ceramic squares (wafers), each containing a single miniaturized component. Components could then be integrated and wired into a bidimensional or tridimensional compact grid. This idea, which seemed very promising in 1957, was proposed to the US Army by Jack Kilby and led to the short-lived Micromodule Program (similar to 1951's Project Tinkertoy). However, as the project was gaining momentum, Kilby came up with a new, revolutionary design: the IC.

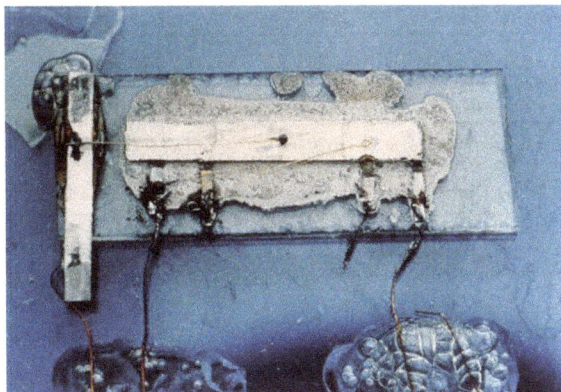

Jack Kilby's original integrated circuit

Newly employed by Texas Instruments, Kilby recorded his initial ideas concerning the integrated circuit in July 1958, successfully demonstrating the first working integrated example on 12 September 1958. In his patent application of 6 February 1959, Kilby described his new device as "a body of semiconductor material ... wherein all the components of the electronic circuit are completely integrated." The first customer for the new invention was the US Air Force.

Kilby won the 2000 Nobel Prize in Physics for his part in the invention of the integrated circuit. His work was named an IEEE Milestone in 2009.

Half a year after Kilby, Robert Noyce at Fairchild Semiconductor developed his own idea of an integrated circuit that solved many practical problems Kilby's had not. Noyce's design was made of silicon, whereas Kilby's chip was made of germanium. Noyce credited Kurt Lehovec of Sprague Electric for the principle of p–n junction isolation caused by the action of a biased p–n junction (the diode) as a key concept behind the IC.

Fairchild Semiconductor was also home of the first silicon-gate IC technology with self-aligned gates, the basis of all modern CMOS computer chips. The technology was developed by Italian physicist Federico Faggin in 1968, who later joined Intel in order to develop the very first single-chip Central Processing Unit (CPU) (Intel 4004), for which he received the National Medal of Technology and Innovation in 2010.

Generations

In the early days of simple integrated circuits, the technology's large scale limited each chip to only a few transistors, and the low degree of integration meant the design process was relatively simple. Manufacturing yields were also quite low by today's standards. As the technology progressed, millions, then billions of transistors could be placed on one chip, and good designs required thorough planning, giving rise to new design methods.

Name	Signification	Year	Transistors number	Logic gates number
SSI	*small-scale integration*	1964	1 to 10	1 to 12
MSI	*medium-scale integration*	1968	10 to 500	13 to 99
LSI	*large-scale integration*	1971	500 to 20,000	100 to 9,999
VLSI	*very large-scale integration*	1980	20,000 to 1,000,000	10,000 to 99,999
ULSI	*ultra-large-scale integration*	1984	1,000,000 and more	100,000 and more

SSI, MSI and LSI

The first integrated circuits contained only a few transistors. Early digital circuits containing tens of transistors provided a few logic gates, and early linear ICs such as the Plessey SL201 or the Philips TAA320 had as few as two transistors. The number of transistors in an integrated circuit has increased dramatically since then. The term "large scale integration" (LSI) was first used by IBM scientist Rolf Landauer when describing the theoretical concept; that term gave rise to the terms "small-scale integration" (SSI), "medium-scale integration" (MSI), "very-large-scale integration" (VLSI), and "ultra-large-scale integration" (ULSI). The early integrated circuits were SSI.

SSI circuits were crucial to early aerospace projects, and aerospace projects helped inspire development of the technology. Both the Minuteman missile and Apollo program needed lightweight digital computers for their inertial guidance systems. Although the Apollo guidance computer led and motivated integrated-circuit technology, it was the Minuteman missile that forced it into mass-production. The Minuteman missile program and various other Navy programs accounted for the total $4 million integrated circuit market in 1962, and by 1968, U.S. Government space and defense spending still accounted for 37% of the $312 million total production. The demand by the U.S. Government supported the nascent integrated circuit market until costs fell enough to allow firms to penetrate the industrial, and eventually, the consumer markets. The average price per integrated circuit dropped from $50.00 in 1962 to $2.33 in 1968. Integrated circuits began to appear in consumer products by the turn of the decade, a typical application being FM inter-carrier sound processing in television receivers.

The first MOS chips were small-scale integration chips for NASA satellites.

The next step in the development of integrated circuits, taken in the late 1960s, introduced devices which contained hundreds of transistors on each chip, called "medium-scale integration" (MSI).

In 1964, Frank Wanlass demonstrated a single-chip 16-bit shift register he designed, with an incredible (at the time) 120 transistors on a single chip.

MSI devices were attractive economically because while they cost little more to produce than SSI devices, they allowed more complex systems to be produced using smaller circuit boards, less assembly work (because of fewer separate components), and a number of other advantages.

Further development, driven by the same economic factors, led to "large-scale integration" (LSI) in the mid-1970s, with tens of thousands of transistors per chip.

SSI and MSI devices often were manufactured by masks created by hand-cutting Rubylith. An engineer would inspect and verify the completeness of each mask. LSI devices contain so many transistors, interconnecting wires, and other features that it is considered impossible for a human to check the masks or even do the original design entirely by hand. The engineer depends on computer programs and other hardware aids to do most of this work.

Integrated circuits such as 1K-bit RAMs, calculator chips, and the first microprocessors, that began to be manufactured in moderate quantities in the early 1970s, had under 4000 transistors. True LSI circuits, approaching 10,000 transistors, began to be produced around 1974, for computer main memories and second-generation microprocessors.

VLSI

The final step in the development process, starting in the 1980s and continuing through the present, was "very-large-scale integration" (VLSI). The development started with hundreds of thousands of transistors in the early 1980s, and continues beyond several billion transistors as of 2009.

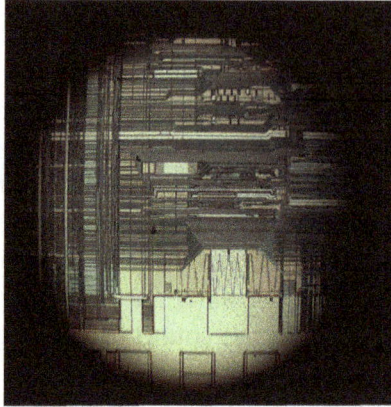

Upper interconnect layers on an Intel 80486DX2 microprocessor die

Multiple developments were required to achieve this increased density. Manufacturers moved to smaller design rules and cleaner fabrication facilities, so that they could make chips with more transistors and maintain adequate yield. The path of process improvements was summarized by the International Technology Roadmap for Semiconductors (ITRS). Design tools improved enough to make it practical to finish these designs in a reasonable time. The more energy-efficient CMOS replaced NMOS and PMOS, avoiding a prohibitive increase in power consumption.

In 1986 the first one-megabit RAM chips were introduced, containing more than one million transistors. Microprocessor chips passed the million-transistor mark in 1989 and the billion-transistor mark in 2005. The trend continues largely unabated, with chips introduced in 2007 containing tens of billions of memory transistors.

ULSI, WSI, SOC and 3D-IC

To reflect further growth of the complexity, the term ULSI that stands for "ultra-large-scale integration" was proposed for chips of more than 1 million transistors.

Wafer-scale integration (WSI) is a means of building very large integrated circuits that uses an entire silicon wafer to produce a single "super-chip". Through a combination of large size and reduced packaging, WSI could lead to dramatically reduced costs for some systems, notably massively parallel supercomputers. The name is taken from the term Very-Large-Scale Integration, the current state of the art when WSI was being developed.

A system-on-a-chip (SoC or SOC) is an integrated circuit in which all the components needed for a computer or other system are included on a single chip. The design of such a device can be complex and costly, and building disparate components on a single piece of silicon may compromise the efficiency of some elements. However, these drawbacks are offset by lower manufacturing and assembly costs and by a greatly reduced power budget: because signals among the components are kept on-die, much less power is required.

A three-dimensional integrated circuit (3D-IC) has two or more layers of active electronic components that are integrated both vertically and horizontally into a single circuit. Communication between layers uses on-die signaling, so power consumption is much lower than in equivalent separate circuits. Judicious use of short vertical wires can substantially reduce overall wire length for faster operation.

Advances in Integrated Circuits

ICs have consistently migrated to smaller feature sizes over the years, allowing more circuitry to be packed on each chip. This increased capacity per unit area can be used to decrease cost or increase functionality, in its modern interpretation, states that the number of transistors in an integrated circuit doubles every two years. In general, as the feature size shrinks, almost everything improves—the cost per transistor and the switching power consumption per transistor go down, and the speed goes up. However, ICs with nanometerscale devices are not without their problems, principal among which is leakage current, although innovations in high-κ dielectrics aim to solve these problems. Since these speed and power consumption gains are apparent to the end user, there is fierce competition among the manufacturers to use finer geometries. This process, and the expected progress over the next few years, is described by the International Technology Roadmap for Semiconductors (ITRS).

The die from an Intel 8742, an 8-bit microcontroller that includes a CPU running at 12 MHz, 128 bytes of RAM, 2048 bytes of EPROM, and I/O in the same chip

Among the most advanced integrated circuits are the microprocessors or "cores", which control everything from computers and cellular phones to digital microwave ovens. Digital memory chips and application-specific integrated circuits (ASICs) are examples of other families of integrated circuits that are important to the modern information society. While the cost of designing and developing a complex integrated circuit is quite high, when spread across typically millions of production units the individual IC cost is minimized. The performance of ICs is high because the small size allows short traces which in turn allows low power logic (such as CMOS) to be used at fast switching speeds.

In current research projects, integrated circuits are also developed for sensoric applications in medical implants or other bioelectronic devices. Special sealing techniques have to be applied in such biogenic environments to avoid corrosion or biodegradation of the exposed semiconductor materials. As one of the few materials well established in CMOS technology, titanium nitride (TiN) turned out as exceptionally stable and well suited for electrode applications in medical implants.

Computer Assisted Design

Classification

Integrated circuits can be classified into analog, digital and mixed signal (both analog and digital on the same chip).

A CMOS 4511 IC in a DIP

Digital integrated circuits can contain anywhere from one to millions of logic gates, flip-flops, multiplexers, and other circuits in a few square millimeters. The small size of these circuits allows high speed, low power dissipation, and reduced manufacturing cost compared with board-level integration. These digital ICs, typically microprocessors, DSPs, and microcontrollers, work using binary mathematics to process "one" and "zero" signals.

Analog ICs, such as sensors, power management circuits, and operational amplifiers, work by processing continuous signals. They perform functions like amplification, active filtering, demodulation, and mixing. Analog ICs ease the burden on circuit designers by having expertly designed analog circuits available instead of designing a difficult analog circuit from scratch.

ICs can also combine analog and digital circuits on a single chip to create functions such as A/D converters and D/A converters. Such mixed-signal circuits offer smaller size and lower cost, but must carefully account for signal interference.

Modern electronic component distributors often further sub-categorize the huge variety of integrated circuits now available:

- Digital ICs are further sub-categorized as logic ICs, memory chips, interface ICs (level shifters, serializer/deserializer, etc.), Power Management ICs, and programmable devices.

- Analog ICs are further sub-categorized as linear ICs and RF ICs.

- mixed-signal integrated circuits are further sub-categorized as data acquisition ICs (including A/D converters, D/A converter, digital potentiometers) and clock/timing ICs.

Manufacturing

Fabrication

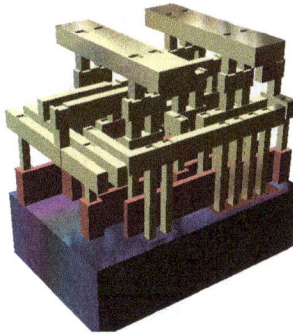

Rendering of a small standard cell with three metal layers (dielectric has been removed). The sand-colored structures are metal interconnect, with the vertical pillars being contacts, typically plugs of tungsten. The reddish structures are polysilicon gates, and the solid at the bottom is the crystalline silicon bulk.

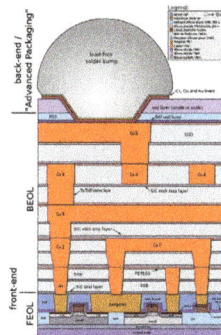

Schematic structure of a CMOS chip, as built in the early 2000s. The graphic shows LDD-MISFET's on an SOI substrate with five metallization layers and solder bump for flip-chip bonding. It also shows the section for FEOL (front-end of line), BEOL (back-end of line) and first parts of back-end process.

The semiconductors of the periodic table of the chemical elements were identified as the most likely materials for a solid-state vacuum tube. Starting with copper oxide, proceeding to germanium, then silicon, the materials were systematically studied in the 1940s and 1950s. Today, monocrystalline silicon is the main substrate used for ICs although some III-V compounds of the periodic table such as gallium arsenide are used for specialized applications like LEDs, lasers, solar cells and the highest-speed integrated circuits. It took decades to perfect methods of creating crystals without defects in the crystalline structure of the semiconducting material.

Semiconductor ICs are fabricated in a planar process which includes three key process steps – imaging, deposition and etching. The main process steps are supplemented by doping and cleaning.

Mono-crystal silicon wafers (or for special applications, silicon on sapphire or gallium arsenide wafers) are used as the substrate. Photolithography is used to mark different areas of the substrate to be doped or to have polysilicon, insulators or metal (typically aluminium) tracks deposited on them.

- Integrated circuits are composed of many overlapping layers, each defined by photolithography, and normally shown in different colors. Some layers mark where various dopants are diffused into the substrate (called diffusion layers), some define where additional ions are implanted (implant layers), some define the conductors (polysilicon or metal layers), and some define the connections between the conducting layers (via or contact layers). All components are constructed from a specific combination of these layers.

- In a self-aligned CMOS process, a transistor is formed wherever the gate layer (polysilicon or metal) crosses a diffusion layer.

- Capacitive structures, in form very much like the parallel conducting plates of a traditional electrical capacitor, are formed according to the area of the "plates", with insulating material between the plates. Capacitors of a wide range of sizes are common on ICs.

- Meandering stripes of varying lengths are sometimes used to form on-chip resistors, though most logic circuits do not need any resistors. The ratio of the length of the resistive structure to its width, combined with its sheet resistivity, determines the resistance.

- More rarely, inductive structures can be built as tiny on-chip coils, or simulated by gyrators.

Since a CMOS device only draws current on the transition between logic states, CMOS devices consume much less current than bipolar devices.

A random access memory is the most regular type of integrated circuit; the highest density devices are thus memories; but even a microprocessor will have memory on the chip. Although the structures are intricate – with widths which have been shrinking for decades – the layers remain much thinner than the device widths. The layers of material are fabricated much like a photographic process, although light waves in the visible spectrum cannot be used to "expose" a layer of material, as they would be too large for the features. Thus photons of higher frequencies (typically ultraviolet) are used to create the patterns for each layer. Because each feature is so small, electron microscopes are essential tools for a process engineer who might be debugging a fabrication process.

Each device is tested before packaging using automated test equipment (ATE), in a process known as wafer testing, or wafer probing. The wafer is then cut into rectangular blocks, each of which is called a die. Each good die (plural dice, dies, or die) is then connected into a package using aluminium (or gold) bond wires which are thermosonically bonded to pads, usually found around the edge of the die. . Thermosonic bonding was first introduced by A. Coucoulas which provided a reliable means of forming these vital electrical connections to the outside world. After packaging, the devices go through final testing on the same or similar ATE used during wafer probing. Industrial CT scanning can also be used. Test cost can account for over 25% of the cost of fabrication on lower-cost products, but can be negligible on low-yielding, larger, or higher-cost devices.

As of 2005, a fabrication facility (commonly known as a semiconductor fab) costs over US$1 billion to construct. The cost of a fabrication facility rises over time (Rock's law) because much of the operation is automated. Today, the most advanced processes employ the following techniques:

- The wafers are up to 300 mm in diameter (wider than a common dinner plate).

- Use of 32 nanometer or smaller chip manufacturing process. Intel, IBM, NEC, and AMD are using ~32 nanometers for their CPU chips. IBM and AMD introduced immersion lithography for their 45 nm processes

- Copper interconnects where copper wiring replaces aluminium for interconnects.

- Low-K dielectric insulators.

- Silicon on insulator (SOI).

- Strained silicon in a process used by IBM known as strained silicon directly on insulator (SSDOI).

- Multigate devices such as tri-gate transistors being manufactured by Intel from 2011 in their 22 nm process.

Packaging

The earliest integrated circuits were packaged in ceramic flat packs, which continued to be used by the military for their reliability and small size for many years. Commercial circuit packaging quickly moved to the dual in-line package (DIP), first in ceramic and later in plastic. In the 1980s pin counts of VLSI circuits exceeded the practical limit for DIP packaging, leading to pin grid array (PGA) and leadless chip carrier (LCC) packages. Surface mount packaging appeared in the early 1980s and became popular in the late 1980s, using finer lead pitch with leads formed as either gull-wing or J-lead, as exemplified by small-outline integrated circuit – a carrier which occupies an area about 30–50% less than an equivalent DIP, with a typical thickness that is 70% less. This package has "gull wing" leads protruding from the two long sides and a lead spacing of 0.050 inches.

A Soviet MSI nMOS chip made in 1977, part of a four-chip calculator set designed in 1970

In the late 1990s, plastic quad flat pack (PQFP) and thin small-outline package (TSOP) packages became the most common for high pin count devices, though PGA packages are still often used for high-end microprocessors. Intel and AMD are currently transitioning from PGA packages on high-end microprocessors to land grid array (LGA) packages.

Ball grid array (BGA) packages have existed since the 1970s. Flip-chip Ball Grid Array packages, which allow for much higher pin count than other package types, were developed in the 1990s. In an FCBGA package the die is mounted upside-down (flipped) and connects to the package balls via a package substrate that is similar to a printed-circuit board rather than by wires. FCBGA packages allow an array of input-output signals (called Area-I/O) to be distributed over the entire die rather than being confined to the die periphery.

Traces out of the die, through the package, and into the printed circuit board have very different electrical properties, compared to on-chip signals. They require special design techniques and need much more electric power than signals confined to the chip itself.

When multiple dies are put in one package, it is called SiP, for System In Package. When multiple dies are combined on a small substrate, often ceramic, it's called an MCM, or Multi-Chip Module. The distinction between a big MCM and a small printed circuit board is sometimes fuzzy.

Chip Labeling and Manufacture Date

Most integrated circuits large enough to include identifying information include four common sections: the manufacturer's name or logo, the part number, a part production batch number and serial number, and a four-digit code that identifies when the chip was manufactured. Extremely small surface mount technology parts often bear only a number used in a manufacturer's lookup table to find the chip characteristics.

The manufacturing date is commonly represented as a two-digit year followed by a two-digit week code, such that a part bearing the code 8341 was manufactured in week 41 of 1983, or approximately in October 1983.

Intellectual Property

The possibility of copying by photographing each layer of an integrated circuit and preparing photomasks for its production on the basis of the photographs obtained is the main reason for the introduction of legislation for the protection of layout-designs. The Semiconductor Chip Protection Act (SCPA) of 1984 established a new type of intellectual property protection for mask works that are fixed in semiconductor chips. It did so by amending title 17 of the United States chapter 9

A diplomatic conference was held at Washington, D.C., in 1989, which adopted a Treaty on Intellectual Property in Respect of Integrated Circuits (IPIC Treaty).

The Treaty on Intellectual Property in respect of Integrated Circuits, also called Washington Treaty or IPIC Treaty (signed at Washington on 26 May 1989) is currently not in force, but was partially integrated into the TRIPS agreement.

National laws protecting IC layout designs have been adopted in a number of countries.

Other Developments

In the 1980s, programmable logic devices were developed. These devices contain circuits whose logical function and connectivity can be programmed by the user, rather than being fixed by the integrated circuit manufacturer. This allows a single chip to be programmed to implement different LSI-type functions such as logic gates, adders and registers. Current devices called field-programmable gate arrays can now implement tens of thousands of LSI circuits in parallel and operate up to 1.5 GHz.

The techniques perfected by the integrated circuits industry over the last three decades have been used to create very small mechanical devices driven by electricity using a technology known as microelectromechanical systems. These devices are used in a variety of commercial and military applications. Example commercial applications include DLP projectors, inkjet printers, and accelerometers and MEMS gyroscopes used to deploy automobile airbags.

As of 2014, the vast majority of all transistors are fabricated in a single layer on one side of a chip of silicon in a flat 2-dimensional planar process. Researchers have produced prototypes of several promising alternatives, such as:

- fabricating transistors over the entire surface of a small sphere of silicon.

- various approaches to stacking several layers of transistors to make a three-dimensional integrated circuit, such as through-silicon via, "monolithic 3D", stacked wire bonding, etc.

- transistors built from other materials: graphene transistors, molybdenite transistors, carbon nanotube field-effect transistor, gallium nitride transistor, transistor-like nanowire electronic devices, organic field-effect transistor, etc.

- modifications to the substrate, typically to make "flexible transistors" for a flexible display or other flexible electronics, possibly leading to a roll-away computer.

In the past, radios could not be fabricated in the same low-cost processes as microprocessors. But since 1998, a large number of radio chips have been developed using CMOS processes. Examples include Intel's DECT cordless phone, or Atheros's 802.11 card.

Future developments seem to follow the multi-core multi-microprocessor paradigm, already used by the Intel and AMD dual-core processors. Rapport Inc. and IBM started shipping the KC256 in 2006, a 256-core microprocessor. Intel, as recently as February–August 2011, unveiled a prototype, "not for commercial sale" chip that bears 80 cores. Each core is capable of handling its own task independently of the others. This is in response to the heat-versus-speed limit that is about to be reached using existing transistor technology. This design provides a new challenge to chip programming. Parallel programming languages such as the open-source X10 program-ming language are designed to assist with this task.

Since the early 2000s, the integration of optical functionality (optical computing) into silicon chips has been actively pursued in both academic research and in industry resulting in the successful commercialization of silicon based integrated optical transceivers combining optical devices (modulators, detectors, routing) with CMOS based electronics.

Silicon Labelling and Graffiti

To allow identification during production most silicon chips will have a serial number in one corner. It is also common to add the manufacturer's logo. Ever since ICs were created, some chip designers have used the silicon surface area for surreptitious, non-functional images or words. These are sometimes referred to as chip art, silicon art, silicon graffiti or silicon doodling.

ICS and IC Families

- The 555 timer IC

- The 741 operational amplifier

- 7400 series TTL logic building blocks

- 4000 series, the CMOS counterpart to the 7400 series

- Intel 4004, the world's first microprocessor, which led to the famous 8080 CPU and then the IBM PC's 8088, 80286, 486 etc.

- The MOS Technology 6502 and Zilog Z80 microprocessors, used in many home computers of the early 1980s

- The Motorola 6800 series of computer-related chips, leading to the 68000 and 88000 series (used in some Apple computers and in the 1980s Commodore Amiga series).

- The LM-series of analog integrated circuits.

Modulation

In electronics and telecommunications, modulation is the process of varying one or more properties of a periodic waveform, called the carrier signal, with a modulating signal that typically contains information to be transmitted.

In telecommunications, modulation is the process of conveying a message signal, for example a digital bit stream or an analog audio signal, inside another signal that can be physically transmitted. Modulation of a sine waveform transforms a baseband message signal into a passband signal.

A modulator is a device that performs modulation. A demodulator (sometimes detector or demod) is a device that performs demodulation, the inverse of modulation. A modem (from modulator–demodulator) can perform both operations.

The aim of analog modulation is to transfer an analog baseband (or lowpass) signal, for example an audio signal or TV signal, over an analog bandpass channel at a different frequency, for example over a limited radio frequency band or a cable TV network channel.

The aim of digital modulation is to transfer a digital bit stream over an analog bandpass channel, for example over the public switched telephone network (where a bandpass filter limits the frequency range to 300–3400 Hz) or over a limited radio frequency band.

Analog and digital modulation facilitate frequency division multiplexing (FDM), where several low pass information signals are transferred simultaneously over the same shared physical medium, using separate passband channels (several different carrier frequencies).

The aim of digital baseband modulation methods, also known as line coding, is to transfer a digital bit stream over a baseband channel, typically a non-filtered copper wire such as a serial bus or a wired local area network.

The aim of pulse modulation methods is to transfer a narrowband analog signal, for example a phone call over a wideband baseband channel or, in some of the schemes, as a bit stream over another digital transmission system.

In music synthesizers, modulation may be used to synthesise waveforms with an extensive overtone spectrum using a small number of oscillators. In this case the carrier frequency is typically in the same order or much lower than the modulating waveform.

Analog Modulation Methods

In analog modulation, the modulation is applied continuously in response to the analog information signal.

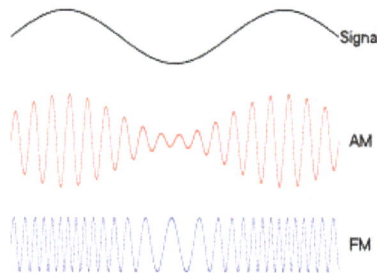

A low-frequency message signal (top) may be carried by an AM or FM radio wave.

List of Analog Modulation Techniques

Common analog modulation techniques are:

- Amplitude modulation (AM) (here the amplitude of the carrier signal is varied in accordance to the instantaneous amplitude of the modulating signal)

    - Double-sideband modulation (DSB)

        - Double-sideband modulation with carrier (DSB-WC) (used on the AM radio broadcasting band)

        - Double-sideband suppressed-carrier transmission (DSB-SC)

- Double-sideband reduced carrier transmission (DSB-RC)

- Single-sideband modulation (SSB, or SSB-AM)

  - Single-sideband modulation with carrier (SSB-WC)

  - Single-sideband modulation suppressed carrier modulation (SSB-SC)

  o Vestigial sideband modulation (VSB, or VSB-AM)

  o Quadrature amplitude modulation (QAM)

- Angle modulation, which is approximately constant envelope

  o Frequency modulation (FM) (here the frequency of the carrier signal is varied in accordance to the instantaneous amplitude of the modulating signal)

  o Phase modulation (PM) (here the phase shift of the carrier signal is varied in accordance with the instantaneous amplitude of the modulating signal)

  o Transpositional Modulation (TM), in which the waveform inflection is modified resulting in a signal where each quarter cycle is transposed in the modulation process. TM is a pesudo-analog modulation (AM). Where an AM carrier also carries a phase variable phase f($\acute{o}$). TM is f(AM,$\acute{o}$)

Digital Modulation Methods

In digital modulation, an analog carrier signal is modulated by a discrete signal. Digital modulation methods can be considered as digital-to-analog conversion, and the corresponding demodulation or detection as analog-to-digital conversion. The changes in the carrier signal are chosen from a finite number of M alternative symbols (the modulation alphabet).

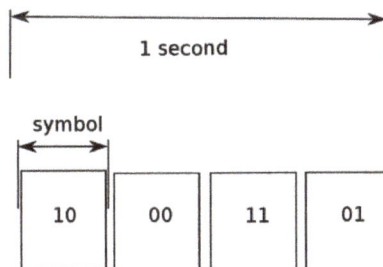

Schematic of 4 baud (8 bit/s) data link containing arbitrarily chosen values.

A simple example: A telephone line is designed for transferring audible sounds, for example tones, and not digital bits (zeros and ones). Computers may however communicate over a telephone line by means of modems, which are representing the digital bits by tones, called symbols. If there are four alternative symbols (corresponding to a musical instrument that can generate four different tones, one at a time), the first symbol may represent the bit sequence 00, the second 01, the third 10 and the fourth 11. If the modem plays a melody consisting of 1000 tones per second, the symbol rate is 1000 symbols/second, or baud. Since each tone (i.e., symbol) represents a message consisting of two digital bits in this example, the bit rate is twice the symbol rate, i.e. 2000 bits per second. This is similar to the technique used by dialup modems as opposed to DSL modems.

According to one definition of digital signal, the modulated signal is a digital signal. According to another definition, the modulation is a form of digital-to-analog conversion. Most textbooks would consider digital modulation schemes as a form of digital transmission, synonymous to data transmission; very few would consider it as analog transmission.

Fundamental Digital Modulation Methods

The most fundamental digital modulation techniques are based on keying:

- PSK (phase-shift keying): a finite number of phases are used.

- FSK (frequency-shift keying): a finite number of frequencies are used.

- ASK (amplitude-shift keying): a finite number of amplitudes are used.

- QAM (quadrature amplitude modulation): a finite number of at least two phases and at least two amplitudes are used.

In QAM, an inphase signal (or I, with one example being a cosine waveform) and a quadrature phase signal (or Q, with an example being a sine wave) are amplitude modulated with a finite number of amplitudes, and then summed. It can be seen as a two-channel system, each channel using ASK. The resulting signal is equivalent to a combination of PSK and ASK.

In all of the above methods, each of these phases, frequencies or amplitudes are assigned a unique pattern of binary bits. Usually, each phase, frequency or amplitude encodes an equal number of bits. This number of bits comprises the symbol that is represented by the particular phase, frequency or amplitude.

If the alphabet consists of $M = 2^N$ alternative symbols, each symbol represents a message consisting of N bits. If the symbol rate (also known as the baud rate) is f_S symbols/second (or baud), the data rate is Nf_S bit/second.

For example, with an alphabet consisting of 16 alternative symbols, each symbol represents 4 bits. Thus, the data rate is four times the baud rate.

In the case of PSK, ASK or QAM, where the carrier frequency of the modulated signal is constant, the modulation alphabet is often conveniently represented on a constellation diagram, showing the amplitude of the I signal at the x-axis, and the amplitude of the Q signal at the y-axis, for each symbol.

Modulator and Detector Principles of Operation

PSK and ASK, and sometimes also FSK, are often generated and detected using the principle of QAM. The I and Q signals can be combined into a complex-valued signal I+jQ (where j is the imaginary unit). The resulting so called equivalent lowpass signal or equivalent baseband signal is a complex-valued representation of the real-valued modulated physical signal (the so-called passband signal or RF signal).

These are the general steps used by the modulator to transmit data:

1. Group the incoming data bits into codewords, one for each symbol that will be transmitted.

2. Map the codewords to attributes, for example amplitudes of the I and Q signals (the equivalent low pass signal), or frequency or phase values.

3. Adapt pulse shaping or some other filtering to limit the bandwidth and form the spectrum of the equivalent low pass signal, typically using digital signal processing.

4. Perform digital to analog conversion (DAC) of the I and Q signals (since today all of the above is normally achieved using digital signal processing, DSP).

5. Generate a high frequency sine carrier waveform, and perhaps also a cosine quadrature component. Carry out the modulation, for example by multiplying the sine and cosine waveform with the I and Q signals, resulting in the equivalent low pass signal being frequency shifted to the modulated passband signal or RF signal. Sometimes this is achieved using DSP technology, for example direct digital synthesis using a waveform table, instead of analog signal processing. In that case the above DAC step should be done after this step.

6. Amplification and analog bandpass filtering to avoid harmonic distortion and periodic spectrum.

At the receiver side, the demodulator typically performs:

1. Bandpass filtering.

2. Automatic gain control, AGC (to compensate for attenuation, for example fading).

3. Frequency shifting of the RF signal to the equivalent baseband I and Q signals, or to an intermediate frequency (IF) signal, by multiplying the RF signal with a

local oscillator sinewave and cosine wave frequency.

4. Sampling and analog-to-digital conversion (ADC) (sometimes before or instead of the above point, for example by means of undersampling).

5. Equalization filtering, for example a matched filter, compensation for multipath propagation, time spreading, phase distortion and frequency selective fading, to avoid intersymbol interference and symbol distortion.

6. Detection of the amplitudes of the I and Q signals, or the frequency or phase of the IF signal.

7. Quantization of the amplitudes, frequencies or phases to the nearest allowed symbol values.

8. Mapping of the quantized amplitudes, frequencies or phases to codewords (bit groups).

9. Parallel-to-serial conversion of the codewords into a bit stream.

10. Pass the resultant bit stream on for further processing such as removal of any error-correcting codes.

As is common to all digital communication systems, the design of both the modulator and demodulator must be done simultaneously. Digital modulation schemes are possible because the transmitter-receiver pair have prior knowledge of how data is encoded and represented in the communications system. In all digital communication systems, both the modulator at the transmitter and the demodulator at the receiver are structured so that they perform inverse operations.

Non-coherent modulation methods do not require a receiver reference clock signal that is phase synchronized with the sender carrier signal. In this case, modulation symbols (rather than bits, characters, or data packets) are asynchronously transferred. The opposite is coherent modulation.

List of Common Digital Modulation Techniques

The most common digital modulation techniques are:

- Phase-shift keying (PSK)

    o Binary PSK (BPSK), using M=2 symbols

    o Quadrature PSK (QPSK), using M=4 symbols

    o 8PSK, using M=8 symbols

    o 16PSK, using M=16 symbols

- o Differential PSK (DPSK)
- o Differential QPSK (DQPSK)
- o Offset QPSK (OQPSK)
- o π/4–QPSK
- Frequency-shift keying (FSK)
  - o Audio frequency-shift keying (AFSK)
  - o Multi-frequency shift keying (M-ary FSK or MFSK)
  - o Dual-tone multi-frequency (DTMF)
- Amplitude-shift keying (ASK)
- On-off keying (OOK), the most common ASK form
  - o M-ary vestigial sideband modulation, for example 8VSB
- Quadrature amplitude modulation (QAM), a combination of PSK and ASK
  - o Polar modulation like QAM a combination of PSK and ASK
- Continuous phase modulation (CPM) methods
  - o Minimum-shift keying (MSK)
  - o Gaussian minimum-shift keying (GMSK)
  - o Continuous-phase frequency-shift keying (CPFSK)
- Orthogonal frequency-division multiplexing (OFDM) modulation
  - o Discrete multitone (DMT), including adaptive modulation and bit-loading
- Wavelet modulation
- Trellis coded modulation (TCM), also known as Trellis modulation
- Spread-spectrum techniques
  - o Direct-sequence spread spectrum (DSSS)
  - o Chirp spread spectrum (CSS) according to IEEE 802.15.4a CSS uses pseudo-stochastic coding
  - o Frequency-hopping spread spectrum (FHSS) applies a special scheme for channel release

MSK and GMSK are particular cases of continuous phase modulation. Indeed, MSK is a particular case of the sub-family of CPM known as continuous-phase frequency-shift keying (CPFSK) which is defined by a rectangular frequency pulse (i.e. a linearly increasing phase pulse) of one symbol-time duration (total response signaling).

OFDM is based on the idea of frequency-division multiplexing (FDM), but the multiplexed streams are all parts of a single original stream. The bit stream is split into several parallel data streams, each transferred over its own sub-carrier using some conventional digital modulation scheme. The modulated sub-carriers are summed to form an OFDM signal. This dividing and recombining helps with handling channel impairments. OFDM is considered as a modulation technique rather than a multiplex technique, since it transfers one bit stream over one communication channel using one sequence of so-called OFDM symbols. OFDM can be extended to multi-user channel access method in the orthogonal frequency-division multiple access (OFDMA) and multi-carrier code division multiple access (MC-CDMA) schemes, allowing several users to share the same physical medium by giving different sub-carriers or spreading codes to different users.

Of the two kinds of RF power amplifier, switching amplifiers (Class D amplifiers) cost less and use less battery power than linear amplifiers of the same output power. However, they only work with relatively constant-amplitude-modulation signals such as angle modulation (FSK or PSK) and CDMA, but not with QAM and OFDM. Nevertheless, even though switching amplifiers are completely unsuitable for normal QAM constellations, often the QAM modulation principle are used to drive switching amplifiers with these FM and other waveforms, and sometimes QAM demodulators are used to receive the signals put out by these switching amplifiers.

Automatic Digital Modulation Recognition (ADMR)

Automatic digital modulation recognition in intelligent communication systems is one of the most important issues in software defined radio and cognitive radio. According to incremental expanse of intelligent receivers, automatic modulation recognition becomes a challenging topic in telecommunication systems and computer engineering. Such systems have many civil and military applications. Moreover, blind recognition of modulation type is an important problem in commercial systems, especially in software defined radio. Usually in such systems, there are some extra information for system configuration, but considering blind approaches in intelligent receivers, we can reduce information overload and increase transmission performance. Obviously, with no knowledge of the transmitted data and many unknown parameters at the receiver, such as the signal power, carrier frequency and phase offsets, timing information, etc., blind identification of the modulation is a difficult task. This becomes even more challenging in real-world scenarios with multipath fading, frequency-selective and time-varying channels.

There are two main approaches to automatic modulation recognition. The first approach uses likelihood-based methods to assign an input signal to a proper class. Another recent approach is based on feature extraction.

Digital Baseband Modulation or Line Coding

The term digital baseband modulation (or digital baseband transmission) is synonymous to line codes. These are methods to transfer a digital bit stream over an analog baseband channel (a.k.a. lowpass channel) using a pulse train, i.e. a discrete number of signal levels, by directly modulating the voltage or current on a cable. Common examples are unipolar, non-return-to-zero (NRZ), Manchester and alternate mark inversion (AMI) codings.

Pulse Modulation Methods

Pulse modulation schemes aim at transferring a narrowband analog signal over an analog baseband channel as a two-level signal by modulating a pulse wave. Some pulse modulation schemes also allow the narrowband analog signal to be transferred as a digital signal (i.e., as a quantized discrete-time signal) with a fixed bit rate, which can be transferred over an underlying digital transmission system, for example, some line code. These are not modulation schemes in the conventional sense since they are not channel coding schemes, but should be considered as source coding schemes, and in some cases analog-to-digital conversion techniques.

Demodulation

Demodulation is extracting the original information-bearing signal from a modulated carrier wave. A demodulator is an electronic circuit (or computer program in a software-defined radio) that is used to recover the information content from the modulated carrier wave. There are many types of modulation so there are many types of demodulators. The signal output from a demodulator may represent sound (an analog audio signal), images (an analog video signal) or binary data (a digital signal).

These terms are traditionally used in connection with radio receivers, but many other systems use many kinds of demodulators. For example, in a modem, which is a contraction of the terms modulator/demodulator, a demodulator is used to extract a serial digital data stream from a carrier signal which is used to carry it through a telephone line, coaxial cable, or optical fiber.

History

Demodulation was first used in radio receivers. In the wireless telegraphy radio systems used during the first 3 decades of radio (1884-1914) the transmitter did not communicate audio (sound) but transmitted information in the form of pulses of radio waves that represented text messages in Morse code. Therefore, the receiver merely had to detect the presence or absence of the radio signal, and produce a click sound. The device that did this was called a detector. The first detectors were coherers, simple devices that acted as a switch. The term detector stuck, was used for other types of demodulators and continues to be used to the present day for a demodulator in a radio receiver.

The first type of modulation used to transmit sound over radio waves was amplitude modulation (AM), invented by Reginald Fessendon around 1900. An AM radio signal can be demodulated by rectifying it, removing the radio frequency pulses on one side of the carrier, converting it from alternating current (AC) to a pulsating direct current (DC). The amplitude of the DC varies with the modulating audio signal, so it can drive an earphone. Fessendon invented the first AM demodulator in 1904 called the electrolytic detector, consisting of a short needle dipping into a cup of dilute acid. The same year John Ambrose Fleming invented the Fleming valve or thermionic diode which could also rectify an AM signal.

Techniques

There are several ways of demodulation depending on how parameters of the baseband signal such as amplitude, frequency or phase are transmitted in the carrier signal. For example, for a signal modulated with a linear modulation like AM (amplitude modulation), we can use a synchronous detector. On the other hand, for a signal modulated with an angular modulation, we must use an FM (frequency modulation) demodulator or a PM (phase modulation) demodulator. Different kinds of circuits perform these functions.

Many techniques such as carrier recovery, clock recovery, bit slip, frame synchronization, rake receiver, pulse compression, Received Signal Strength Indication, error detection and correction, etc., are only performed by demodulators, although any specific demodulator may perform only some or none of these techniques.

Many things can act as a demodulator, if they pass the radio waves on nonlinearly. For example, near a powerful radio station, it has been known for the metal sides of a van to demodulate the radio signal as sound.

AM Radio

An AM signal encodes the information onto the carrier wave by varying its amplitude in direct sympathy with the analogue signal to be sent. There are two methods used to demodulate AM signals:

- The envelope detector is a very simple method of demodulation that does not require a coherent demodulator. It consists of an envelope detector that can be a rectifier (anything that will pass current in one direction only) or other non-linear that enhances one half of the received signal over the other and a low-pass filter. The rectifier may be in the form of a single diode or may be more complex. Many natural substances exhibit this rectification behaviour, which is why it was the earliest modulation and demodulation technique used in radio. The filter is usually an RC low-pass type but the filter function can sometimes be achieved by relying on the limited frequency response of the circuitry following the rectifier. The crystal set exploits the simplicity of AM modulation to produce a receiver with very few parts, using the crystal as the rectifier and the limited frequency response of the headphones as the filter.

- The product detector multiplies the incoming signal by the signal of a local oscillator with the same frequency and phase as the carrier of the incoming signal. After filtering, the original audio signal will result.

SSB is a form of AM in which the carrier is reduced or suppressed entirely, which require coherent demodulation.

FM Radio

Frequency modulation (FM) has numerous advantages over AM such as better fidelity and noise immunity. However, it is much more complex to both modulate and demodulate a carrier wave with FM and AM predates it by several decades.

There are several common types of FM demodulators:

- The quadrature detector, which phase shifts the signal by 90 degrees and multiplies it with the unshifted version. One of the terms that drops out from this operation is the original information signal, which is selected and amplified.

- The signal is fed into a PLL and the error signal is used as the demodulated signal.

- The most common is a Foster-Seeley discriminator. This is composed of an electronic filter which decreases the amplitude of some frequencies relative to others, followed by an AM demodulator. If the filter response changes linearly with frequency, the final analog output will be proportional to the input frequency, as desired.

- A variant of the Foster-Seeley discriminator called the ratio detector

- Another method uses two AM demodulators, one tuned to the high end of the band and the other to the low end, and feed the outputs into a difference amplifier.

- Using a digital signal processor, as used in software-defined radio.

References

- Winston, Brian (1998). Media Technology and Society: A History : From the Telegraph to the Internet. Routledge. p. 221. ISBN 978-0-415-14230-4.

- Mindell, David A. (2008). Digital Apollo: Human and Machine in Spaceflight. The MIT Press. ISBN 978-0-262-13497-2.

- Ginzberg, Eli (1976). Economic impact of large public programs: the NASA Experience. Olympus Publishing Company. p. 57. ISBN 0-913420-68-9.

- Bob Johnstone (1999). We were burning: Japanese entrepreneurs and the forging of the electronic age. Basic Books. pp. 47–48. ISBN 978-0-465-09118-8.

- Ke-Lin Du & M. N. S. Swamy (2010). Wireless Communication Systems: From RF Subsystems to 4G Enabling Technologies. Cambridge University Press. p. 188. ISBN 978-0-521-11403-5.

- Topol, A.W.; Tulipe, D.C.La; Shi, L; et., al. "Three-dimensional integrated circuits". ieee.org. International Business Machines Corporation (IBM). Retrieved 21 September 2014.

- "Milestones:First Semiconductor Integrated Circuit (IC), 1958". IEEE Global History Network. IEEE. Retrieved 3 August 2011.

4

Various Tags in Radio Frequency Identification

This chapter deals exclusively with the tracking tags used in radio frequency identification. The various types discussed are clipped tag, bag tag, tochatag and ear tag. Each tag has a different design, composition, function and application. The reader is informed about these differentiating characteristics and the latest advances in each. The topics discussed in the chapter are of great importance to broaden the existing knowledge on radio frequency identification.

Clipped Tag

The clipped tag is a radio frequency identification tag designed to enhance consumer privacy. Radio frequency identification or RFID is an identification technology in which information stored in semiconductor chips contained in RFID tags is communicated by means of radio waves to RFID readers. The most simple passive RFID tags do not have batteries or transmitters. They get their energy from the field of the reader. They transfer their information to the reader by modulating the signal that is reflected back to the reader by the tag. Because tags depend on the reader for power their range is limited, typically up to 10 meters or 30 feet for UHF RFID tags.

Today, the public uses RFID tags for many applications including electronic toll collection, E-ZPass for example, or the Speedpass which is used as a credit token for the purchase of gasoline. The retail supply chain uses RFID tags to monitor the passage of pallets and cases at loading dock doors. The expectation for the future is for RFID tags to be used for the labelling of items for retail sale. Concerns for individual privacy have been raised because the RFID tags may be read by invisible radio waves without the knowledge of the holder of the tagged item.

The privacy-protecting RFID tag, the "clipped tag" has been suggested by IBM. The clipped tag puts the option of privacy protection in the hands of the consumer. After the point of sale, a consumer may tear off a portion of the tag, much like the way in which a ketchup packet is opened. This allows the transformation of a long-range tag into a proximity tag that still may be read, but only at short range – less than a few inches or centimeters. The modification of the tag may be confirmed visually. The tag may still be used later for returns, recalls, or recycling. The clipped tag was listed among the Wall Street Journal Technology Innovation Winners for 2006. Two US patents were issued for this invention in 2007.

Other mechanisms designed to protect privacy for RFID item tagging for retail use are the EPCglobal kill command and the RSA blocker tag.

Clipped Tag Development

The concept of the clipped tag was first introduced in a paper authored by IBM researchers Paul Moskowitz and Guenter Karjoth in 2005, RFID Journal, November 7, 2005. In their paper, presented at the 2005 ACM Workshop on Privacy in the Electronic Society, the authors suggest that by providing the consumer with a means to shorten the antenna, the read range of the tag may be reduced from many meters to just a few centimeters. Several mechanisms were suggested. The mechanisms included perforating the tag like a sheet of postage stamps to allow the tearing off of a portion of the antenna. Another proposed mechanism was to manufacture the tag antenna with exposed conducting lines which could be scratched off by the consumer.

IBM teamed up with Marnlen RFiD, a manufacturer of RFID labels, and Printronix, a maker of RFID printers, to demonstrate prototypes of the Clipped Tag, Wired News, May 1, 2006. The tag took the form of a garment hang tag with v-shaped notches in the edges and perforations to direct the tearing of the tag. Reactions by RFID privacy experts were favorable to the invention. According to Wired, Robert Atkinson, president of the Information Technology and Innovation Foundation, said "The Clipped Tag shows that IBM is addressing privacy concerns, even those that are unreasonable." Subsequently, IBM and Marnlen RFiD announced that Marnlen had licensed the technology from IBM and was shipping samples to select users, RFID Update, November, 2006.

Bag Tag

Example of IATA airport code printed on a baggage tag, showing DCA (Ronald Reagan Washington National Airport).

Bag tags, also known as baggage tags, baggage checks or luggage tickets, have traditionally been used by bus, train and airline companies to route passenger luggage that is checked on to the final destination. The passenger stub is typically handed to the passenger or attached to the ticket envelope: a) to aid the passenger in identifying their bag among similar bags at the destination baggage carousel; b) as proof—still requested at a few airports—that the passenger is not removing someone else's bag from the baggage reclaim hall; c) as a means for the passenger and carrier to identify and trace a specific bag that has gone astray and was not delivered at the destination.

The carriers' liability is restricted to published tariffs and international agreements.

History

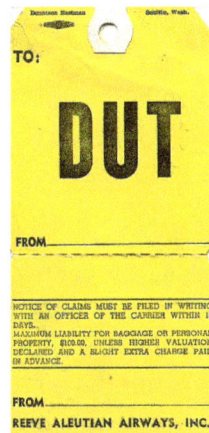

Bag tag for a 1972 flight to Unalaska Airport on Reeve Aleutian Airways

Invention

The first "separable coupon ticket" was patented by John Michael Lyons of Moncton, New Brunswick on June 5, 1882. The ticket showed the issuing station, the destination and a consecutive number for reference. The lower half of the ticket was given to the passenger, while the upper half, with a hole at the top, was inserted into a brass sleeve and then attached to the baggage by a strap.

At some point, reinforced paper tags were introduced. They are designed to not detach as easily as older tags during transport.

Warsaw Convention

The Warsaw Convention of 1929, specifically article 4, established the criteria for issuing a baggage check or luggage ticket. This agreement also established limit of liability on checked baggage.

Previous Bag Tags

Prior to the 1990s, airline bag tags consisted of a paper tag attached with a string.

The tag contained basic information:

- Airline/carrier name

- flight number

- baggage tag number, composed of the two letter airline code and six digits

- destination airport code

These tags became obsolete as they offered little security and were easy to replicate.

Current Bag Tags

Current bag tags include a bar code using the Interleaved 2 of 5 symbology. These bag tags are printed using thermal or barcode printer on an adhesive thermal paper stock. This printed strip is then attached to the luggage at check in. This allows for automated sorting of the bags by bar code readers.

There are two ways that bar code baggage tags are read; hand held scanners and in-line arrays. In-line arrays are built into the baggage conveyor system and use a 360 degree array of lasers to read the bar code tags from multiple angles as baggage and the orientation of the bar code tag can shift as it goes through the conveyor belt system. One of the limitations of this system is that to read bar codes from the bottom of the belt laser arrays are placed below the gap between two sections of conveyor belt. Due to the frequent build-up of debris and dust on these lower arrays the rate of successful reads can be low. Frequently, the "read rate", the percent of bar code tags successful read by these arrays can be as low as 85%. This means that more than one out of ten bar code baggage tags are not successfully read and these bags are shunted off for manual reading resulting in extra labour and delay.

For flights departing from an international airport within the European Union, bag tags are issued with green edges. Passengers are eligible to take these bags through a separate "Blue Channel" at Customs.

Bar codes can not be automatically scanned without direct sight and undamaged print. Forced by reading problems with poorly printed, obscured, crumpled, scored or otherwise damaged bar codes, some airlines have started using radio-frequency identification (RFID) chips embedded in the tags. In the US, McCarran International Airport has installed an RFID system throughout the airport. Hong Kong International Airport has also installed an RFID system. The International Air Transport Association (IATA) is working to standardize RFID bag tags.

British Airways are currently trialling the use of re-usable electronic luggage tags featuring electronic paper technology. The passenger has to check-in using the British Airways smartphone app, then the passenger holds their phone close the tag and it will transmit the flight details and barcode to the tag using NFC technology. As the tag utilises electronic paper, the battery only has to power the tag when the passenger is sending the data to the tag.

FastTrack Company have developed a re-usable electronic luggage tag product called the Eviate eTag. This is also electronic paper-based, but is not limited to a single airline. The passenger will check-in using a supported airline's smartphone app, then is able to send the relevant flight information to the tag via Bluetooth Low Energy.

Qantas introduced Q Bag Tags in 2011. Unlike the British Airways tags, they do not feature a screen which means there is no barcode to scan. This has limited the use of the tags to domestic flights within Australia on the Qantas network. The tags were initially given free of charge to members of the Qantas Frequent Flyer program with Silver, Gold or Platinum status. The tags can also be purchased for A$29.95.

Identification

The first automated baggage sorting systems were developed by Eastern Air Lines in the 1980s, first at their Miami International Airport hub. Other airlines soon followed with their own systems including United Air Lines, TWA, Delta and American Airlines. In the early days none of these systems were interchangeable. In some systems the bar code was used to represent a three letter destination airport code, in others a two digit sorting symbol telling the system which pier to deliver the bag to.

As a result of the bombing of Air India Flight 182 on June 23 1985 the airline industry, led by IATA, convened the Baggage Security Working Group (BSWG) to change the international standards to require passenger baggage reconciliation. The Chairman of the BSWG John Vermilye, of Eastern Airlines, proposed that the industry adopt the already proven License Plate system. This concept used the bar code to represent the baggage tag number and at check-in this number was associated with the passenger details including flight number, destination, connection information and could even include class of service to indicate priority handling. Working with Allen Davidson of Litton Industries with whom Eastern developed the license plate concept, the BSWG adopted the license plate concept as the common industry standard for passenger baggage reconciliation. Initially the bar code or license plate was used to facilitate matching baggage with passengers to ensure that only baggage of passengers who have boarded a flight are carried on the aircraft. This standard was adopted in IATA Resolution in 1987. By 1989 the license plate concept was expanded to become the industry standard for automated baggage sorting as well and at this time the bar codes were enlarged to facilitate automated reading. The bar code is shown in two different orientations or in a "T" shape called the "orthogonal" representation.

The term license plate is the official term used by the IATA, the airlines, and the airports for the 10-digit numeric code on a bag tag issued by a carrier or handling agent at check-in.

The license plate is printed on the carrier tag in bar code form and in human-readable form, as defined in Resolution 740 in the IATA Passenger Services Conference Resolutions Manual (published annually by IATA). Each digit in a license plate has a specific meaning. The license plate is an index number linking a bag to a Baggage Source Message (BSM) sent by a carrier's departure control system to an airport's baggage handling system. It is the message that contains the flight details and passenger information, thus enabling an automated baggage handling system to sort a bag automatically once it has scanned the bar code on the carrier tag. Thus these two things are essential for automated sorting of baggage. Note that the human-readable license plate may contain a 2-character IATA carrier code instead of an IATA 3-digit carrier code. For example, BA728359 instead of 0125728359, but the bar code will always be the full 10 digits (0125728359 in the example - 125 and BA being, respectively, the IATA 3-digit code and IATA 2-character code for British Airways). The first digit of a 10-digit license plate is not part of the carrier code. It can be in the range 0 to 9: 0 for interline or online tags, 1 for fallback tags (pre-printed or demand-printed tags only for use by the local baggage handling system if it cannot receive BSMs from a carrier's departure control system due to a fault in the latter or in communication between it and the baggage handling system, as defined in IATA Recommended Practice 1740b) and 2 for Rush tags. The purpose of numbers in the range 3 to 9 as the first digit of the 10-digit license plate is undefined by IATA but can be used by each carrier for its specific needs (commonly used as a million indicator for the normal 6-digit tag number).

Besides the license plate number, the tag also has:

- Name of airport of arrival

- Departure time

- IATA airport code of airport of arrival

- Airline code and flight number

- Name of passenger identified with the baggage (last name, first name)

Touchatag

Touchatag (previously TikiTag) was an RFID service for consumers, application developers and operators/enterprises created by Alcatel-Lucent. Consumers could use RFID tags to trigger what touchatag called Applications, which could include opening a webpage, sending a text message, shutting down the computer, or running a custom application created through the software's API, via the application developer network. Touchatag applications were also compatible with NFC enabled phones like the Nokia 6212. TikiTag was launched as an Open Beta on October 1, 2008. And it was rebranded to touchatag on February 15, 2009. Touchatag also sold RFID hardware, like a starter package with 1 USB RFID reader and 10 RFID tags (stickers), for which the client software was compatible with Windows XP and Vista, along with Mac OS X 10.4 and up. Touchatag was carried by Amazon.com, ThinkGeek, Firebox.com and getDigital.de along with Touchatag's own Online Store. Touchatag also marketed their products' underlying technology for enterprise and operator solutions. Touchatag announced an agreement with Belgacom PingPing on jointly developing the contactless market and announced a commercial pilot with Accor Services. On June 27, 2012 the Touchatag team has announced the shutdown of the project. inviting users to use IOTOPE "a similar open source Internet Of Things service" which itself has no apparent activity since November 2012.

Service

Touchatag's core offering was the touchatag service, based on the "application correlation service" and allowed tag, reader and application management. For consumers, the web interface allowed to link RFID tags (and 2D barcode tags, more precisely QR Code) to applications. Application developers could use the correlation API to use the ACS functionalities to create contactless applications. For businesses, this ACS was extended with an RFID/NFC tag and reader catalogue, and applications like loyalty, interactive advertising and couponing.

Hardware

The reader provided was an ACR122U Tag Reader, from Advanced Card Systems. The tags shipped with the reader were MiFare Ultralight tags.

Software Support

Official Clients

Touchatag hardware was supported by its makers on Microsoft Windows and Mac OS X platforms, and required registration on the website to work. An unsupported application was also available for Linux platforms. Like the Mac OS X application, the Linux application used PCSC-Lite for hardware access.

Unofficial Clients

- TagEventor (no apparent activity since 2011) is an open-source client from the Autelic (dead link) association that works on Linux platforms, and does not use the web service from touchatag. It uses the PCSC-Lite daemon and can be run in foreground or daemon mode to make tag events available to user-space applications.

Programming Libraries

- touchatag-processing is a Processing library from Augusto Esteves that allows users to connect and read from multiple touchatag readers on the Windows platform. This library works with simply the reader's drivers, so there's no need to install any official or unofficial clients.

Ear Tag

An ear tag is a plastic or metal object used for identification of domestic livestock and other animals. If the ear tag uses Radio Frequency Identification Device (RFID) technology it is referred to as an electronic ear tag. Electronic ear tags conform to international standards ISO 11784 and ISO 11785 working at 134.2 kHz, as well as ISO/IEC 18000-6C operating in the UHF spectrum. There are other non-standard systems such as Destron working at 125 kHz. Although there are many shapes of ear tags, the main types in current use are as follows:

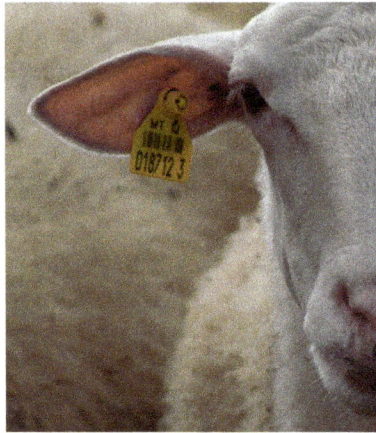

A sheep with an ear tag.

- Flag-shaped ear tag: two discs joined through the ear, one or both bearing a wide, flat plastic surface on which identification details are written or printed in large, easily legible script.

- Button-shaped ear tag: two discs joined through the ear.

- Plastic clip ear tag: a moulded plastic strip, folded over the edge of the ear and joined through it.

- Metal ear tag: an aluminium, steel or brass rectangle with sharp points, clipped over the edge of the ear, with the identification stamped into it.

Each of these except the metal type may carry a RFID chip, which normally carries an electronic version of the same identification number.

Overview

An ear tag usually carries an Animal Identification Number (AIN) or code for the animal, or for its herd or flock. Non electronic ear tags may be simply handwritten for the convenience of the farmer (these are known as "management tags"). Alternatively this identification number (ID) may be assigned by an organisation, such as the Meat and

Livestock Association (MLA), which is a not-for-profit organisation owned by cattle, sheep and goat producers; funded by a levy on livestock sales with Federal Government input. Electronic tags may also show other information about the animal, including other related identification numbers; such as the Property Identification Code (PIC) for the properties the animals have been located. In the case of MLA's NLIS the movement of certain species of livestock (primarily cattle, goats and sheep) must be recorded in the online database within 24 hours of the movement; and include the PICs of the properties the animals are travelling between. The National Livestock Identification System (NLIS) of Australia regulations require that all cattle be fitted with a RFID device in the form of an ear tag or rumen bolus (a cylindrical object placed in the rumen) before movement from the property and that the movement be reported to the NLIS. However, if animals are tagged for internal purposes in a herd or farm, IDs need not be unique in larger scales. The NLIS now also requires sheep and goats to use an ear tag that has the Property Identification Code inscribed on it. These ear tags and boluses are complemented by transport documents supplied by vendors that are used for identification and tracking. A similar system is used for cattle in the European Union (EU), each bovine animal having a passport document and tag in each ear carrying the same number. Sheep and goats in the EU have a tag in one or both ears carrying the official number of their flock and also for breeding stock an individual number for each animal; one of these tags (usually the left) must have a RFID chip (or the chip may instead be carried in a rumen bolus or on an anklet).

An ear tag can be applied with an ear tag applicator, however there are also specially-designed tags that can be applied by hand. Depending on the purpose of the tagging, an animal may be tagged on one ear or both. There may be requirements for the placement of ear tags, and care must be taken to ensure they are not placed too close to the edge of the ear pinnae; which may leave the tag vulnerable to being ripped out accidentally. If there exists a national animal identification programme in a country, animals may be tagged on both ears for the sake of increased security and effectiveness, or as a legal requirement. If animals are tagged for private purposes, usually one ear is tagged. Australian sheep and goats are required to have visually readable ear tags printed with a Property Identification Code (PIC). They are complemented by movement documents supplied by consignors that are used for identification and tracking.

Very small ear tags are available for laboratory animals such as mice and rats. They are usually sold with a device that pierces the animal's ear and installs the tag at the same time. Lab animals can also be identified by other methods such as ear punching or marking, implanted RFID tags (mice are too small to wear an ear tag containing an RFID chip), and dye.

History

Livestock ear tags were developed in 1799 under the direction of Sir Joseph Banks, President of the Royal Society, for identification of Merino sheep in the flock established for King George III. Matthew Boulton designed and produced the first batch of sheep eartags, and produced subsequent batches, modified according to suggestions received from Banks. The first tags were made of tin.

A sow polar bear with ear tag

Ear tags were incorporated as breed identification in the United States with the forming of the International Ohio Improved Chester Association as early as 1895, and stipulated in the Articles of Incorporation, as an association animal and breed identification, of the improved Chester White.

A tagged Black-tailed Prairie Dog

Although ear tags were developed in Canada as early as 1913 as a means to identify cattle when testing for tuberculosis, the significant increase of use of ear tags appeared with the outbreak of BSE in UK. Today, ear tags in a variety of designs are used throughout the world on many species of animal to ensure traceability, to help prevent theft and to control disease outbreaks.

The first ear tags were primarily steel with nickel plating. After World War II, larger, flag-like, plastic tags were developed in the United States. Designed to be visible from a distance, these were applied by cutting a slit in the ear and slipping the arrow-shaped head of the tag through it so that the flag would hang from the ear.

In 1953, the first two-piece, self-piercing plastic ear tag was developed and patented. This tag, which combined the easy application of metal tags with the visibility and colour options of plastic tags, also limited the transfer of blood-borne diseases between animals during the application process.

Some cattle ear tags contain chemicals to control insects such as buffalo fly etc. Metal ear tags are used to identify the date of regulation shearing of stud and show sheep. Today, a large number of manufacturers are in competition for the identification of world livestock population .

In 2004, the U.S. Government asked farmers to use EID or Electronic Identification ear tags on all their cattle. This request was part of the National Animal Identification System (NAIS) spurred by the discovery of the first case of mad cow disease in the United States. Due to poor performance and concern that other people could access their confidential information, only about 30 percent of cattle producers in the United States tried using EID tags using standards based on the low frequency standards, while the UHF standards are being mandated for use in Brazil, Paraguay, and Korea . The United States Department of Agriculture maintains a list of manufacturers approved to sell ear tags in the USA.

Ear tags (conventional and electronic) are used in the EU as official ID system for cattle, sheep and goat, in some cases combined with RFID devices

The International Committee for Animal Recording (ICAR) controls the issue electronic tag numbers under ISO regulation 11784.

The National Livestock Identification System (NLIS) is Australia's system for tracing cattle, sheep and goats from birth to slaughter.

In Canada, the Health of Animals Regulations require approved ear tags on all bison, cattle and sheep that leave the farm of origin, except that a bison or bovine may be moved, without a tag, from the farm of origin to a tagging site. RFID (radio frequency identification) tags are used for cattle in Canada and metal as well as RFID tags have been in use for sheep. Mandatory RFID tagging of sheep in Canada (which was previously scheduled to take effect January 1, 2013) will be deferred to some later date.

Other Forms of Animal Identification

Pigs, cattle and sheep are frequently earmarked with pliers that notch registered owner and/or age marks into the ear. Mares on large horse breeding farms have a plastic tag attached to a neck strap for identification; which preserves their ears free of notches. Dairy cows are sometimes identified with ratchet fastened plastic anklets fitted on the pastern for ready inspection during milking; however NLIS requirements apply to cattle - including both dairy and beef animals. More commonly coloured electrical tape is used as short term ankle identifiers for dairy animals to identify when one teat should not be milked for any reason. Laboratory rodents are often marked with ear tags, ear notches or implantable microchips.

The National Livestock Identification System (NLIS) Australia, formerly used cattle tail tags for property identification and hormone usage declaration.

References

- International Ohio Improved Chester Record Association (c. 1895). "Ear tags". original from Cornell University. Himrods, N.Y. p. 41. Retrieved October 15, 2014.

- "SupportedReadersAndTags - tageventor - Project Hosting on Google Code". Code.google.com. 2009-10-24. Retrieved 2010-04-02.

- "Welcome to the business site | Touchatag Business". Business.touchatag.com. Retrieved 2010-04-02.

5

Applications and Uses of Radio Frequency Identification

The tracking abilities and ease of use makes radio frequency identification find application in numerous fields. Some of the applications discussed are biometric passport, electronic article surveillance, contactless payment, machine-readable passport, smart-dust, transponder timing, telecommunication, intelligent transportation system and transponder. Tracking tags also find application in animal conservation efforts where it is implanted into target species to keep a track of their numbers and whereabouts.

Biometric Passport

A biometric passport, also known as an e-passport, ePassport or a digital passport, is a combined paper and electronic passport that contains biometric information that can be used to authenticate the identity of travelers. It uses contactless smart card technology, including a microprocessor chip (computer chip) and antenna (for both power to the chip and communication) embedded in the front or back cover, or center page, of the passport. Document and chip characteristics are documented in the International Civil Aviation Organization's (ICAO) Doc 9303. The passport's critical information is both printed on the data page of the passport and stored in the chip. Public Key Infrastructure (PKI) is used to authenticate the data stored electronically in the passport chip making it expensive and difficult to forge when all security mechanisms are fully and correctly implemented.

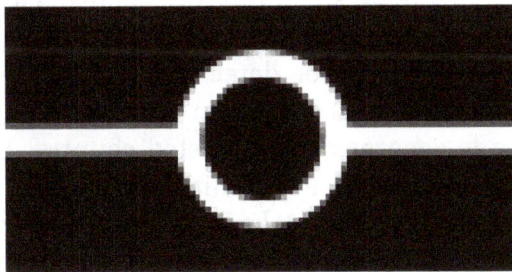

This symbol for biometrics is usually printed on the cover of such passports.

The currently standardized biometrics used for this type of identification system are facial recognition, fingerprint recognition, and iris recognition. These were adopted after assessment of several different kinds of biometrics including retinal scan. The

ICAO defines the biometric file formats and communication protocols to be used in passports. Only the digital image (usually in JPEG or JPEG2000 format) of each biometric feature is actually stored in the chip. The comparison of biometric features is performed outside the passport chip by electronic border control systems (e-borders). To store biometric data on the contactless chip, it includes a minimum of 32 kilobytes of EEPROM storage memory, and runs on an interface in accordance with the ISO/IEC 14443 international standard, amongst others. These standards intend interoperability between different countries and different manufacturers of passport books.

Some national identity cards (for example in the Netherlands, Albania and Brazil) are fully ICAO9303 compliant biometric travel documents. However others, such as the United States Passport Card, are not.

Data Protection

Biometric passports are equipped with protection mechanisms to avoid and/or detect attacks:

- Non-traceable chip characteristics. Random chip identifiers reply to each request with a different chip number. This prevents tracing of passport chips. Using random identification numbers is optional.

- Basic Access Control (BAC). BAC protects the communication channel between the chip and the reader by encrypting transmitted information. Before data can be read from a chip, the reader needs to provide a key which is derived from the Machine Readable Zone: the date of birth, the date of expiry and the document number. If BAC is used, an attacker cannot (easily) eavesdrop transferred information without knowing the correct key. Using BAC is optional.

- Passive Authentication (PA). PA is aimed at identifying modification of passport chip data. The chip contains a file (SOD) that stores hash values of all files stored in the chip (picture, fingerprint, etc.) and a digital signature of these hashes. The digital signature is made using a document signing key which itself is signed by a country signing key. If a file in the chip (e.g. the picture) is changed, this can be detected since the hash value is incorrect. Readers need access to all used public country keys to check whether the digital signature is generated by a trusted country. Using PA is mandatory. According to a September 2011 United States Central Intelligence Agency document released by Wikileaks in December 2014, "Although falsified e-passports will not have the correct digital signature, inspectors may not detect the fraud if the passports are from countries that do not participate in the International Civil Aviation Organization's Public Key Directory (ICAO PKD). Only 15 of over 60 e-passport-issuing countries belong to the PKD program, as of December 2010"

- Active Authentication (AA). AA prevents cloning of passport chips. The chip contains a private key that cannot be read or copied, but its existence can easily be proven. Using AA is optional.

- Extended Access Control (EAC). EAC adds functionality to check the authenticity of both the chip (chip authentication) and the reader (terminal authentication). Furthermore, it uses stronger encryption than BAC. EAC is typically used to protect fingerprints and iris scans. Using EAC is optional. In the European Union, using EAC is mandatory for all documents issued starting 28 June 2009.

- Supplemental Access Control (SAC) was introduced by ICAO in 2009 for addressing BAC weaknesses. It was introduced as a supplement to BAC (for keeping compatibility), but will replace it in the future.

- Shielding the chip. This prevents unauthorized reading. Some countries – including at least the US – have integrated a very thin metal mesh into the passport's cover to act as a shield when the passport cover is closed. The use of shielding is optional.

Inspection Process

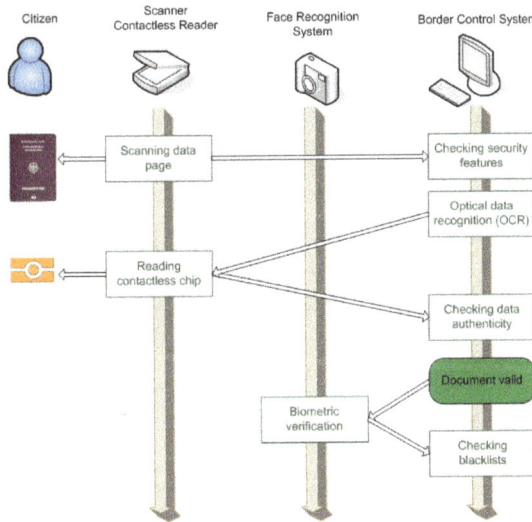

The typical work-flow of an automatic border control system (eGate)

Attacks

Since the introduction of biometric passports several attacks have been presented and demonstrated:

- Non-traceable chip characteristics. In 2008 a Radboud/Lausitz University team demonstrated that it's possible to determine which country a passport chip is from without knowing the key required for reading it. The team fingerprinted error messages of passport chips from different countries. The resulting lookup table allows an attacker to determine from where a chip originated. In 2010 Tom Chothia and Vitaliy Smirnov documented an attack that allows an individual passport to be traced, by sending specific BAC authentication requests.

- Basic Access Control (BAC). In 2005 Marc Witteman showed that the document numbers of Dutch passports were predictable, allowing an attacker to guess/ crack the key required for reading the chip. In 2006 Adam Laurie wrote software that tries all known passport keys within a given range, thus implementing one of Witteman's attacks. Using online flight booking sites, flight coupons and other public information it's possible to significantly reduce the number of possible keys. Laurie demonstrated the attack by reading the passport chip of a Daily Mail's reporter in its envelope without opening it. Note that in some early biometric passports BAC wasn't used at all, allowing attacker to read the chip's content without providing a key.

- Passive Authentication (PA). In 2006 Lukas Grunwald demonstrated that it is trivial to copy passport data from a passport chip into a standard ISO/IEC 14443 smartcard using a standard contactless card interface and a simple file transfer tool. Grunwald used a passport that did not use Active Authentication (anti-cloning) and did not change the data held on the copied chip, thus keeping its cryptographic signature valid. In 2008 Jeroen van Beek demonstrated that not all passport inspection systems check the cryptographic signature of a passport chip. For his demonstration Van Beek altered chip information and signed it using his own document signing key of a non-existing country. This can only be detected by checking the country signing keys that are used to sign the document signing keys. To check country signing keys the ICAO PKD can be used. Only 5 out of 60+ countries are using this central database. Van Beek did not update the original passport chip: instead an ePassport emulator was used. Also in 2008, The Hacker's Choice implemented all attacks and published code to verify the results. The release included a video clip that demonstrated problems by using a forged Elvis Presley passport that is recognized as a valid US passport.

- Active Authentication (AA). In 2005 Marc Witteman showed that the secret Active Authentication key can be retrieved using power analysis. This may allow an attacker to clone passport chips that use the optional Active Authentication anti-cloning mechanism on chips – if the chip design is susceptible to this attack. In 2008 Jeroen van Beek demonstrated that optional security mechanisms can be disabled by removing their presence from the passport index file. This allows an attacker to remove – amongst others – anti-cloning mechanisms (Active Authentication). The attack is documented in supplement 7 of Doc 9303 (R1-p1_v2_sIV_0006) and can be solved by patching inspection system software. Note that supplement 7 features vulnerable examples in the same document that – when implemented – result in a vulnerable inspection process.

- Extended Access Control (EAC). In 2007 Luks Grunwald presented an attack that can make EAC-enabled passport chips unusable. Grunwald states that if an EAC-key – required for reading fingerprints and updating certificates – is stolen or compromised, an attacker can upload a false certificate with an issue date far in the future. The affected chips block read access until the future date is reached.

Opposition

Privacy proponents in many countries question and protest the lack of information about exactly what the passports' chip will contain, and whether they impact civil liberties. The main problem they point out is that data on the passports can be transferred with wireless RFID technology, which can become a major vulnerability. Although this could allow ID-check computers to obtain a person's information without a physical connection, it may also allow anyone with the necessary equipment to perform the same task. If the personal information and passport numbers on the chip are not encrypted, the information might wind up in the wrong hands.

On 15 December 2006, the BBC published an article on the British ePassport, citing the above stories and adding that:

> "Nearly every country issuing this passport has a few security experts who are yelling at the top of their lungs and trying to shout out: 'This is not secure. This is not a good idea to use this technology'", citing a specialist who states "It is much too complicated. It is in places done the wrong way round – reading data first, parsing data, interpreting data, then verifying whether it is right. There are lots of technical flaws in it and there are things that have just been forgotten, so it is basically not doing what it is supposed to do. It is supposed to get a higher security level. It is not."

and adding that the Future of Identity in the Information Society (FIDIS) network's research team (a body of IT security experts funded by the European Union) has "also come out against the ePassport scheme... [stating that] European governments have forced a document on its citizens that dramatically decreases security and increases the risk of identity theft."

Most security measures are designed against untrusted citizens (the "provers"), but the scientific security community recently also addressed the threats from untrustworthy verifiers, such as corrupt governmental organizations, or nations using poorly implemented, unsecure electronic systems. New cryptographic solutions such as private biometrics are being proposed to mitigate threats of mass theft of identity. These are under scientific study, but not yet implemented in biometric passports.

Countries Using Biometric Passports

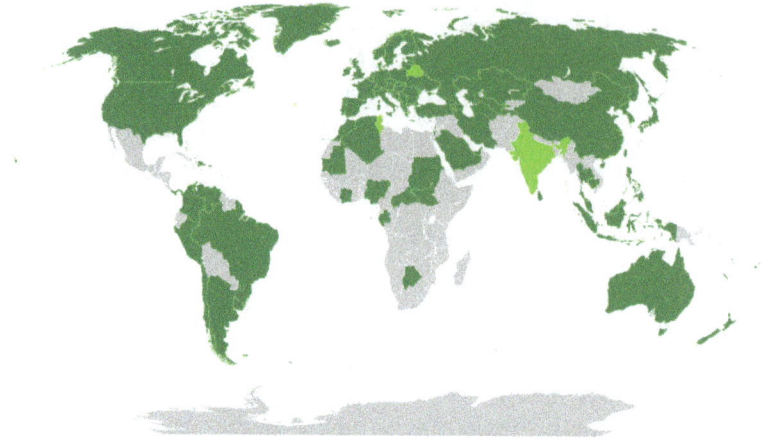

Biometric Passport Map:

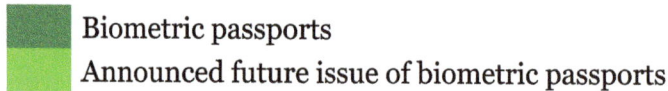

Biometric passports
Announced future issue of biometric passports

European Union

It was planned that, except for Denmark, Ireland and the UK, EU passports would have digital imaging and fingerprint scan biometrics placed on their RFID chips. This combination of biometrics aims to create an unrivaled level of security and protection against fraudulent identification papers. Technical specifications for the new passports have been established by the European Commission. The specifications are binding for the Schengen agreement parties, i.e. the EU countries, except Ireland and the UK, and three of the four European Free Trade Association countries – Iceland, Norway and Switzerland. These countries are obliged to implement machine readable facial images in the passports by 28 August 2006, and fingerprints by 29 June 2009. The European Data Protection Supervisor has stated that the current legal framework fails to "address all the possible and relevant issues triggered by the inherent imperfections of biometric systems". Currently, the British and Irish biometric passports only use a digital image and not fingerprinting. German passports printed after 1 November 2007 contain two fingerprints, one from each hand, in addition to a digital photograph. Romanian passports will also contain two fingerprints, one from each hand. The Netherlands also takes fingerprints and is the only EU member that plans to store these fingerprints centrally. According to EU requirements, only nations that are signatories to the Schengen acquis are required to add fingerprint biometrics.

In the EU nations, passport prices will be

- Austria (available since 16 June 2006): an adult passport costs €75.90, while a chip-free child's version costs €30. As of March 2009 all newly issued adult passports contain fingerprints.

- Belgium (introduced in October 2004): €71 or €41 for children + local taxes. As of May 2014, passports for adults are valid for 7 years.

- Bulgaria (introduced in July 2009; available since 29 March 2010): 40 BGN (€20) for adults. Passports are valid for 5 years.

- Croatia (available since 1 July 2009): 390 HRK (€53). The chip contains two fingerprints and a digital photo of the holder. Since 18 January 2010 only biometric passports can be obtained at issuing offices inside Croatia. Diplomatic missions and consular offices must implement new issuing system until 28 June 2010.

- Cyprus (available since 13 December 2010): €70, valid for 10 years

- Czech Republic (available since 1 September 2006): 600 CZK for adults (valid 10 years), CZK100 for children (valid 5 years). Passports contain fingerprints.

- Denmark (available since 1 August 2006): DKK600 for adults (valid for 10 years), DKK115 for children (valid for 5 years) and DKK350 for over 65 (valid for 10 years). As of January 2012 all newly issued passports contain fingerprints.

- Estonia (available since 22 May 2007): EEK450 (€28.76) (valid for 5 years). As of 29 June 2009, all newly issued passports contain fingerprints.

- Finland (available since 21 August 2006) €53 (valid for up to 5 years). As of 29 June 2009, all newly issued passports contain fingerprints.

- France (available since April 2006): €86 or €89 (depending whether applicant provides photographs), valid for 10 years. As of 16 June 2009, all newly issued passports contain fingerprints.

- Germany (available since November 2005): ≤23-year-old applicants (valid for 6 years) €37.50, >24 years (valid 10 years) €59 Passports issued from 1 November 2007 on include fingerprints.

- Greece (available since 26 August 2006) €84.40 (valid for 5 years). Since June 2009, passports contain fingerprints.

- Hungary (available since 29 August 2006): HUF7,500 (€26), valid for 5 years, HUF14,000 (€48.50) valid for 10 years. As of 29 June 2009, all newly issued passports contain fingerprints.

- Ireland Biometric passport booklets have been available since 16 October 2006, and Biometric passport cards since October 2015.

  32-page passport booklets are priced at €80, 66-page booklets at €110, both valid for 10 years. For children aged between 3 and 18 years the price is €26.50 and the passport booklets are valid for 5 years. Infants' passport booklets for those under 3 years cost €16 and expire 3 years after issue.

Irish biometric passport cards are only available to adults of 18 years and over who already have an Irish passport booklet and cost €35. They expire on the same date as the holder's Irish passport booklet or 5 years after issue, whichever is the shorter period.

(Ireland is not a signatory to the Schengen Acquis and has no obligation or plans to implement fingerprint biometrics)

- Italy (available since 26 October 2006): €116, valid for 10 years. As of January 2010 newly issued passports contain fingerprints.

- Latvia (available since 20 November 2007): an adult passport costs Ls15 (€21.36 [prior to 16, July 2012]), valid for 10 or 5 years.

- Lithuania (available since 28 August 2006): LTL150 (€43). For children up to 16 years old, valid max 5 years. For persons over 16 years old, valid for 10 years.

- Luxembourg (available since 28 August 2006): €30. Valid for 5 years. As of 29 June 2009, all newly issued passports contain fingerprints.

- Malta (available since 8 October 2008): €70 for persons over 16 years old, valid for 10 years, €35 for children between 10–16 years (valid for 5 years) and €14 for children under 10 years (valid for 2 years).

- Netherlands (available since 28 August 2006): Approximately €11 on top of regular passport (€38.33) cost €49.33. Passports issued from 21 September 2009 include fingerprints. Dutch identity cards are lookalike versions of the holder's page of the passport but don't contain fingerprints.

- Poland (available since 28 August 2006): 140PLN (€35) for adults, PLN70 for children aged under 13, free for seniors 70+ years, valid 10 years (5 years for children aged below 13). Passports issued from 29 June 2009 include fingerprints of both index fingers.

- Portugal (available since 31 July 2006 – special passport; 28 August 2006 – ordinary passport): €65 for all citizens valid for 5 years. All passports have 32 pages.

- Romania (available since 31 December 2008): 302 RON (€67), valid for 5 years for those over the age of 6, and for 3 years for those under 6. As of 19 January 2010, new passport includes both facial images and fingerprints.

- Slovakia (available since 15 January 2008): an adult passport (>13 years) costs €33.19 valid for 10 years, while a chip-free child's (5–13 years) version costs €13.27 valid for 5 years and for children under 5 years €8.29, but valid only for 2 years.

- Slovenia (available since 28 August 2006): €42.05 for adults, valid for 10 years. €35.25 for children from 3 to 18 years of age, valid for 5 years. €31.17 for children up to 3 years of age, valid for 3 years. All passports have 32 pages, a 48-page version is available at a €2.50 surcharge. As of 29 June 2009, all newly issued passports contain fingerprints.

- Spain (available since 28 August 2006) at a price of €25 (price at the 22 April 2012). They include fingerprints of both index fingers as of October 2009. (Aged 30 or less a Spanish passport is valid for 5 years, otherwise they remain valid for 10 years).

- Sweden (available since October 2005): SEK 350 (valid for 5 years). As of 1 January 2012, new passport includes both facial images and fingerprints.

- UK (introduced March 2006): £72.50 for adults (valid for 10 years) and £46 for children under the age of 16 (valid for 5 years). (Not Signatory to Schengen Acquis, no obligation to fingerprint biometrics.)

  Unless otherwise noted, none of the issued biometric passports mentioned above include fingerprints as of 5 May 2010.

Albania

The Albanian biometric passport has been available since May 2009, costs 6000 Lekë (€50) and is valid for 10 years. The microchip contains ten fingerprints, the bearer's photo and all the data written on the passport.

Algeria

Algerian biometric passports were introduced on 5 January 2012 with a validity of 10 years for adults.

Argentina

On 15 June 2012, the government announced the availability of a new biometric passport at a cost of 400 pesos, valid for 10 years

Armenia

In July 2012 Armenia introduced two new identity documents to replace ordinary passports of Armenian citizens. One of the documents – ID card with electronic signature and other personal data, is used locally within the country, and the biometric passport with an electronic chip is used for traveling abroad. Electronic chip of biometric passport contains digital images of fingerprints, photo and electronic signature of the passport holder. The passport will be valid for 10 years.

Australia

The Australian biometric passport was introduced in October 2005. The microchip contains the same personal information that is on the colour photo page of the ePassport, including a digitized photograph. A standard (35-Visa Pages) adult passport (>18 years) is A$250 valid for 10 years; for children, the fee is AUD125 valid for 5 years. A Frequent traveler (67-Visa Pages) adult passport (>18 years) is AUD376 valid for 10 years; for children, the fee is AUD188 valid for 5 years. SmartGates have been installed in Australian airports to allow Australian ePassport holders and ePassport holders of several other countries to clear immigration controls more rapidly, and facial recognition technology has been installed at immigration gates.

Azerbaijan

Azerbaijan introduced biometric passports in September 2013 and costing AZN40 (~USD25). The passports will include information about the passport holder's facial features, as well as his finger and palm prints. Each passport will also include a personal identification number. The program covers the development of the appropriate legislative framework and information systems to ensure information security.

Bosnia and Herzegovina

Available since 15 October 2009 and costing 40 KM (€20.51). Valid 10 years for adults and 5 years for younger than 18. Produced by Bundesdruckerei. On 1 June 2010 Bosnia and Herzegovina issued its first EAC passport.

Brazil

Brazil started issuing ICAO compliant passports in December 2006. However just in December 2010 it began to issue passports with microchips, first in the capital Brasília and Goiás state. Since the end of January 2011 this last is available to be issued all over Brazil. It is valid for 5 years for adults and costs R$156.07 (approximately €80). In December 2014, the Federal Police Department extended the validity of the document, from five to ten years.

Brunei

The Bruneian biometric passport was introduced on 17 February 2007. It was produced by German printer Giesecke & Devrient (G&D) following the Visa Waiver Program's requirements. The Bruneian ePassport has the same functions as the other biometric passports.

Cambodia

Cambodia began to issue biometric passports to its citizens on 17 July 2014. The cost for a 5-year passport, issued only to children aged five and under, is 80 USD; while the 10-year passport, issued to all people older than five, costs 100 USD.

Canada

Only the ePassport (Canadian Biometric Passport) is available to Canadians since 1 July 2013. Available for 5 years at a cost of CAD120 or 10 years at CAD160.

Cape Verde

Cape Verde started to issue biometric passports on 26 January 2016. The cost of a biometric passport is 50 euros with a processing time of 30 days. It is noted that the scheme will gradually expand to Cape Verdean diplomatic missions in Boston and Lisbon in the future.

Chile

Chile introduced new biometric passports and national ID cards on 2 September 2013. The newly designed passport booklet has a validity of 5 years.

People's Republic of China

On 30 January 2011, the Ministry of Foreign Affairs of the People's Republic of China launched a trial issuance of e-passports for public affairs. The face, fingerprint and other biometric features of the passport holder will be digitalized and stored in pre-installed contactless smart chip in the passport. On 1 July 2011, the Ministry began issuing biometric passports to all individuals conducting public affairs work overseas on behalf of the Chinese government.

Ordinary biometric passports have been introduced by the Ministry of Public Security starting from 15 May 2012. The cost of a passport is 200 CNY (approximately US$31) for first-time applicants in China and 220 CNY (or US$35) for renewals and passports issued abroad.

Colombia

The Colombian foreign ministry announced that, starting 1 September 2015, new biometric passports will be issued. The only visible change will be that ordinary Colombian passports will now carry the standard biometric symbol at the bottom of the front cover of the booklet.

Dominican Republic

In the Dominican Republic, biometric passports began to be issued in May 2004. However the Dominican biometric passports do not carry the "chip inside" symbol ▄▄. In January 2010, the cost of the passport was 1,250 DOP, about 35–40 USD at that date.

Egypt

The Egyptian Government has, from 5 February 2007, introduced the electronic Passport (e-Passport) and electronic Document of Identity for Visa Purposes (e-Doc/I) which are compliant with the standard of the International Civil Aviation Organization (ICAO). Digital data including holder's personal data and facial image will be contained in the contactless chip embedded in the back cover of e-Passport and e-Doc/I.

Gabon

Available since 23 January 2014. The Gabonese biometric passports carry the "chip inside" symbol (▄▄).

Ghana

Available since 1 March 2010 and costing GH¢ 50.00–100.00 for adults and children. The passports contain several other technological characteristics other than biometric technology. However the Ghanaian biometric passports do not carry the «chip inside» symbol (▄▄), similar to the Pakistani passport, which is mandatory for ICAO-standard electronic passports.

Hong Kong

In 2006, the Immigration Department announced that Unihub Limited (a PCCW subsidiary company heading a consortium of suppliers, including Keycorp) had won the tender to provide the technology to produce biometric passports. In February 2007, the first biometric passport was introduced. The cover of the new biometric passport remains essentially the same as that of previous versions, with the addition of the "electronic passport" logo at the bottom. However, the design of the inner pages has changed substantially. The design conforms with the document design recommendations of the International Civil Aviation Organization. The new ePassport featured in the 2008 Stockholm Challenge Event and was a finalist for the Stockholm Challenge Award in the Public Administration categeory. The Hong Kong SAR ePassport design was praised on account of the "multiple state-of-the-art technologies [which] are seamlessly integrated in the sophisticated Electronic Passport System (e-Passport System)". The cost for a HKSAR passport is HK$370 (or US$48) for a 32-page passport and HK$460 (or US$59) for a 48-page passport.

Iceland

Available since 23 May 2006 and costing ISK5100 (ISK1900 for under 18 and over 67).

India

India has recently initiated first phase deployment of Biometric e-Passport for Diplomatic passport holders in India and abroad. The new passports have been designed indigenously by the Central Passport Organization, the India Security Press, Nashik and IIT Kanpur. The passport contains a security chip with personal data and digital images. Initially, the new passports will have a 64KB chip with a photograph of passport holder and subsequently include the holder's fingerprint(s). The biometric passport has been tested with passport readers abroad and is noted to have a 4-second response time – less than that of a US Passport (10 seconds). The passport need not be carried in a metal jacket for security reasons as it first needs to be passed through a reader, after which generates access keys to unlock the chip data for reader access.

India has also given out a contract to Tata Consultancy Services for issuing e-passports through passport seva kendra. India plans to open 77 such centers across the country to issue these passports.

On 25 June 2008 Indian Passport Authority issued first e-passport to the then President of India, Pratibha Patil. The e-passport is under the first phase of deployment and will be initially restricted to diplomatic passport holders. It is expected to be made available to ordinary citizens from 2013 onwards. The necessary procurements have been initiated by India Security Press, Nasik, installed special machine churning 8 million biometric passports per year in 2010 and the actual transition to the new age passport is expected to begin in the year 2016.

Indonesia

Indonesia started issuing e-Passports on 26 January 2011. The passport costs Rp655,000 (US$66) for the 48-page valid for 5 years, and Rp405,000 (USD41) for the 24-page passport valid for 5 years.

Iran

Iran started issuing biometric diplomatic and service passports in July 2007. Ordinary biometric passports began to be issued on 20 February 2011. The cost of a new passport was approximately US$37 (IRR1,125,000) .

Iraq

Starting February 1, 2010 the Iraqi Ministry of Interior revealed new electronic system to issue the new A series biometric passports in contract with the German SAFE

ID Solutions, the new series is a machine-readable biometric passport available to the public which cost 25,000 dinars or about USD20.

Ireland

On October 16, 2006, the Minister of Foreign Affairs presented the first biometric passports.

Israel

As of July 2013, the Israeli Ministry of the Interior will be issuing biometric passports for those citizens who wish to receive them. For a 2-year pilot period under the Biometric Database Law, this will be optional. As of August 2013, any passport expiring in more than 2 years can be replaced with a biometric one upon request, free of charge. Passports expiring within 2 years will be charged the full fee. The program review that was supposed to be concluded in 2015 was postponed by order of the Minister of the Interior to a later date, due to the controversy regarding the creation of the Biometric Database rather than storing the biometric data only within the passport's chip, as is the practice in most other countries.

Japan

The Japanese government started issuing biometric passports in March 2006. With this, Japan has met requirements under the US Visa Waiver Program which calls for countries to roll out their biometric passports before 26 October 2006.

Kazakhstan

Kazakhstan introduced biometric passport in 2009.

Kosovo

In May 2011, the Ministry of Interior of the Republic of Kosovo announced that biometric passports would be issued in the summer of 2011 after the winning firm is chosen and awarded the production of the passports. The first biometric passports were issued in October 2011.

Lebanon

The Lebanese Directorate General of General Security started issuing biometric passports as of 1 August 2016.

All new Lebanese passports issued are biometric passports and machine-readable contain a contactless smart RFID chip embedded inside the bottom of the front cover under the word "PASSEPORT"

Macau

Applications for electronic passports and electronic travel permits have been started and processed since 1 September 2009.

Macedonia

Available since 2 April 2007 and costs 1500 MKD or c. €22.

Madagascar

The passport is available since 2014 and costs 110,000 Ariary. Since September 2014, it is mandatory for Malagasy citizens to depart the country with a biometric passport.

Malaysia

Malaysia was the first country in the world to issue biometric passports in 1998, after a local company, IRIS Corporation, developed the technology. Malaysia is however not a member of the Visa Waiver Program (VWP) and its first biometric passport did not conform to the same standards as the VWP biometric document because the Malaysian biometric passport was issued several years ahead of the VWP requirement. The difference lies in the storage of fingerprint template instead of fingerprint image in the chip, the rest of the technologies are the same. Also the biometric passport was designed to be read only if the receiving country has the authorization from the Malaysian Immigration Department. Malaysia started issuing ICAO compliant passports from February 2010.

Maldives

Maldives started rolling out its new ePassport to its citizens on 26 July 2006. The new passport follows a completely new design, and features the passport holder's facial and fingerprint information as biometric identifiers. A 32-page Ordinary passport will cost Rf350, while a 64-page Ordinary passport will cost Rf600. Children under the age of 10 years and people applying for passports through diplomatic missions abroad will be issued with a 32-page non-electronic Ordinary passport, which will cost Rf250.

Sovereign Military Order of Malta

Since 2005 the SMOM diplomatic and service passports include biometric features and are compliant with ICAO standards.

Moldova

The Moldovan biometric passport is available from 1 January 2008. The new Moldovan biometric passport costs approximately 760 MDL (€45) and is obligatory from 1 January 2011. The passport of the Republic of Moldova with biometric data contains a chip which holds digital information, including the holder's signature, as well as the traditional information. It is valid for 7 years (for persons over 7) and 4 years (for persons less than 7) respectively. It was introduced as a request of European Union to safeguard the borders between the E.U. and Republic of Moldova.

Montenegro

The Montenegrin biometric passport was introduced in 2008. It costs approximately €40.

Mongolia

The Mongolian ministry of interior stated that first biometric passport will be issue at the end of 2016.

Mauritania

The issuance of the biometric passports was launched 6 May 2011.

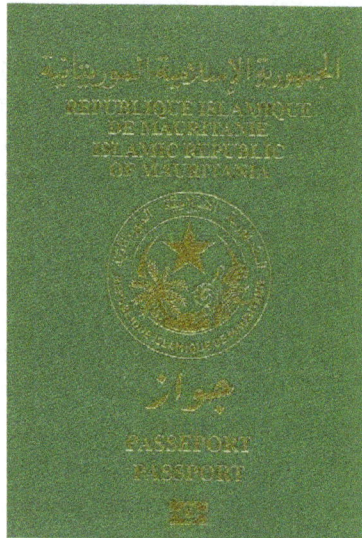

Cover of Mauritanian Biometric Passport

It costs 115.68 US Dollar for issuance and is valid for five years only.

Morocco

The Moroccan biometric passport was introduced in 2008. In December 2009, early limited trials have been extended, and the biometric passport is available from 25 September 2009 to all Moroccan citizens holders of an electronic identity card. It costs 300DH (approximately €27).

Mozambique

Mozambique started to issue biometric passports in September 2014. The issuance of such passports was suspended 40 days later but was resumed in February 2015.

New Zealand

Introduced in November 2005, like Australia and the USA, New Zealand is using the facial biometric identifier. There are two identifying factors: the small symbol on the front cover indicating that an electronic chip has been embedded in the passport, and the polycarbonate leaf in the front (version 2009) of the book inside which the chip is located. The cost is NZ$140 (when applying in person) or NZ$124.50 (when applying online—available only if already holding a passport) for adults, NZ$81.70 for children, valid for five years. However, in 2015 the New Zealand government approved for the reinstatement of a 10-year validity period for passports, which will come into effect on 30 November 2015.

Nigeria

Nigeria is currently one of the few nations in Africa that issues biometric passports, and has done it since 2007.The harmonized ECOWAS Smart electronic passport issued by the Nigerian Immigrations Service is powered by biometric technology in tandem with the International Civil Aviation Organization (ICAO) specifications for international travels.

Travellers' data captured in the biometric passport can be accessed instantly and read by any security agent from any spot of the globe through an integrated network of systems configured and linked to a centrally-coordinated passport data bank managed by the Nigerian Immigrations Service.

Norway

The introduction of biometric passports to Norway began in 2005 and supplied by Setec, costing NOK 450 for adults, or c. €50, NOK 270 for children.

In 2007 the Norwegian government launched a 'multi-modal' biometric enrolment system supplied by Motorola. Motorola's new system enabled multiple public agencies to digitally capture and store fingerprints, 2D facial images and signatures for passports and visas.

The Norwegian biometrics company IDEX ASA has begun development of electronic ID cards (eID) with fingerprint security technology for use throughout the EU.

Pakistan

In 2004, Pakistan became among one of the first countries in the world to issue the biometric passports, which are according to the publisher compliant with ICAO standards and dubbed Multi-biometric e-Passports, however they do not carry the "chip inside" symbol (), which is mandatory for ICAO-standard electronic passports.

As of 2012, Pakistan has adopted the Multi-biometric e-Passport that is now compliant with ICAO standards.

Panama

Panama has issued biometric passports since 2014. The cost of the passport went up from $50 to $100, and the inside contains images of recent government projects.

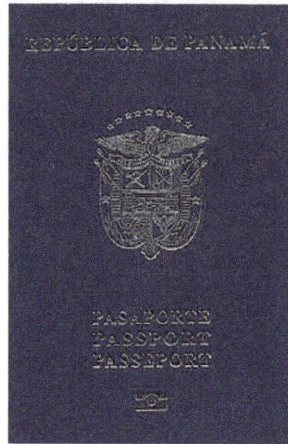

Cover of a Panamanian Biometric Passport

Peru

On 21 February 2016, the Superintendencia Nacional de Migraciones announced that the first Peruvian biometric passports would be delivered by 26 February 2016. The first passport was issued for Peruvian opera singer Juan Diego Flórez. It will feature a new cover, along with several security improvements, in order to be exempted for visas for the Schengen Area. It will cost PEN98.50, approximately USD28, making it the cheapest passport in Latin America.

Philippines

On 11 August 2009, the first biometric passport was issued to then President Gloria Macapagal-Arroyo. The new e-passport has various security features, including a hidden encoded image; an ultra-thin, holographic laminate; and a tamper-proof electronic microchip and is priced at around □950.

Qatar

On 20 April 2008, Qatar started issuing biometric passports which are ICAO compliant. A Qatari passport costs QR200.

Russia

Russian biometric passports were introduced in 2006. As of 2015, they cost 3500 rubles (approximately USD50) and use printed data, photo and fingerprints and are BAC-encrypted. Biometric passports issued after 1 March 2010 are valid for 10 years. Russian biometric passports are currently issued within Russia and in all of its consulates.

From 1 January 2015, the Government of Russia has issued passports which contain fingerprints.

Saudi Arabia

On 21 June 2006, Saudi Arabia started issuing biometric passports which are ICAO compliant. A Saudi Arabian passport costs SR150.

Serbia

Available since 7 July 2008, and cost 3.600 RSD or approximately €32. (Aged 3 or less a Serbian passport is valid for 3 years, aged 3 to 14 it is valid for 5 years, otherwise passport remain valid for 10 years.)

Singapore

The Immigration and Checkpoints Authority (ICA) of Singapore introduced the Singapore biometric passport (BioPass) on 15 August 2006. With this, Singapore has met requirements under the US Visa Waiver Program which calls for countries to roll out their biometric passports before 26 October 2006.

Somalia

The new "e-passport" of Somalia was introduced and approved by the nation's Transitional Federal Government on 10 October 2006. It costs $100 USD to apply for Somalis living inside of Somalia, and $150 USD for Somalis living abroad. Somalia is now the first country on the African continent to have introduced the "e-passport".

South Korea

The Ministry of Foreign Affairs and Trade of South Korea started issuing biometric passports to its citizens on 25 August 2008. The cost is fixed to 55,000 Won or 55 US Dollars, and the validity of ordinary passport is 10 years.

South Sudan

The Republic of South Sudan started issuing internationally recognized electronic passports in January 2012. The passports were officially launched by the President Salva Kiir Mayardit on 3 January 2012 in a ceremony in Juba. The new passport will be valid for five years.

Slovakia

Biometric passports were first issued in Slovakia in 2008. The latest version was issued in 2014 and contains a contactless chip in the biodata card that meets ICAO specifications.

Sri Lanka

From the 10 August 2015, the Department of Immigration and Emigration Sri Lanka has begun issuing ICAO compliant biometric passports to the public.

Sudan

The Republic of the Sudan started issuing electronic passports to citizens in May 2009. The new electronic passport will be issued in three categories. The citizen's passport (ordinary passport) will be issued to ordinary citizens and will contain 48 pages. Business men/women who need to travel often will have a commercial passport that will contain 64 pages. Smaller passports that contain 32 pages only will be issued to children. The microprocessor chip will contain the holder's information. Cost to obtain a new passport will be SDG250 (approximately USD100), SDG200 for students and SDG100 for kids. The validity of the citizen's passport will be five years, or seven years for the commercial passport.

Switzerland

The Swiss biometric passport has been available since 4 September 2006. By a narrow majority of 50.14%, Swiss voters decided in May 2009 to accept the introduction of a biometric passport. Since 1 March 2010, all issued passports are biometric, containing a photograph and two fingerprints recorded electronically. The costs are CHF 140.00 for adults and CHF 60.00 for children (−18 years old).

Taiwan

The Taiwanese biometric passport has been available since 29 December 2008. It costs NT$1,600 for an ordinary passport with either 3, 5 or 10 years validity. Taiwanese Central Engraving and Printing Plant prints passports for the Ministry of Foreign Affairs of Taiwan for several decades. During this period, the passport has been redesigned various times. The current e-passport (or known as biometric passport) is fitted with RFID technology that facilitates Taiwanese passport immigration clearances worldwide.

Tajikistan

Biometric passports will be issued in Tajikistan from 1 February 2010. On 27 August 2009, Tajik Ministry of Foreign Affairs and German Muhlbauer signed a contract on purchase of blank biometric passports and appropriate equipment for Tajikistan.

Thailand

The Ministry of Foreign Affairs of Thailand introduced the first biometric passport for Diplomats and Government officials on 26 May 2005. From 1 June 2005, a limited quantity of 100 passports a day was issued for Thai citizens, however, on 1 August 2005 a full operational service was installed and Thailand became the first country in Asia to issue an ICAO compliant biometric passport.

Togo

In August 2009, Togo became one of the first African countries to introduce biometric passports. The price of the passport was then set at 30,000 CFA Francs for Togolese residing in Togo. For Togolese residing abroad, the price varies.

Tunisia

The Tunisia ministry of interior stated that it will start issuing biometric passports at the end of year 2016.

Turkey

Turkish passports which are compatible with European Union standards have been available since 1 June 2010. Colours of the new biometric passports have also been changed. Accordingly, regular passports; claret red, special passports; bottle green and diplomatic passports wrap black colours.

Most recently Turkish Minister of the State announced that the government is printing the new passports at government minting office since the private contractor failed to deliver.

The current cost of issuing a 10-year passport in Turkey is TRY620.60 (approximately US$215).

Turkmenistan

Turkmenistan became the first country in ex-USSR, in mid-Asia region to issue an ICAO-compliant biometric passport. The passport is available since 10 July 2008.

Ukraine

According to law, Ukraine was supposed to issue biometric passports and identity cards on 1 January 2013. However, they did not become available until two years later in January 2015 and are fully compatible with European Union standards.

United Arab Emirates

The UAE ministry of interior stated that it would start issuing Emirati biometric passports at the end of 2010.

United States

The biometric version of the U.S. passport (sometimes referred to as an electronic passport) has descriptive data and a digitized passport photo on its contactless chips, and does not have fingerprint information placed onto the contactless chip. However, the chip is large enough (64 kilobytes) for inclusion of biometric identifiers. The U.S. Department of State first issued these passports in 2006, and since August 2007 issues biometric passports only. Non-biometric passports are valid until their expiration dates.

Although a system able to perform a facial-recognition match between the bearer and his or her image stored on the contactless chip is desired, it is unclear when such a system will be deployed by the U.S. Department of Homeland Security at its ports of entry.

A high level of security became a priority for the United States after the attacks of 11 September 2001. High security required cracking down on counterfeit passports. In October 2004, the production stages of this high-tech passport commenced as the U.S. Government Printing Office (GPO) issued awards to the top bidders of the program. The awards totaled to roughly $1,000,000 for startup, development, and testing. The driving force of the initiative is the U.S. Enhanced Border Security and Visa Entry Reform Act of 2002 (also known as the "Border Security Act"), which states that such smartcard identity cards will be able to replace visas. As for foreigners travelling to the U.S., if they wish to enter U.S. visa-free under the Visa Waiver Program (VWP), they are now required to possess machine-readable passports that comply with international standards. Additionally, for travellers holding a valid passport issued on or after 26 October 2006, such a passport must be a biometric passport if used to enter the U.S. visa-free under the VWP.

Uruguay

The Uruguayan Ministry of the Interior started to issue biometric passports to Uruguayan citizens on 16 October 2015. The new passport complies with the standards set forth by the Visa Waiver Program of the United States.

Uzbekistan

In Uzbekistan, 23 June 2009 Islam Karimov issued a Presidential Decree "On measures to further improve the passport system in the Republic of Uzbekistan." On 29 December 2009 the President of Uzbekistan signed a decree to change the dates for a phased exchange of populations existing passport to the biometric passport. In accordance with this decree, biometric passports will be phased in, beginning with 1 January 2011. In the first phase, the biometric passport will be issued to employees of ministries, departments and agencies of the republic, individuals who travel abroad or outside the country, as well as citizens who receive a passport in connection with the achievement of a certain age or for other grounds provided by law. The second phase will be for the rest of the population who will be able to get new passports for the period from 2012 to 2015.

Venezuela

Issued after July 2007, Venezuela was the first Latin American country issuing passports including RFID chips along other major security improvements. The chip has photo and fingerprints data.

Animal Identification

Animal identification using a means of marking is a process done to identify and track specific animals. It is done for a variety of reasons including verification of ownership, biosecurity control, and tracking for research or agricultural purposes.

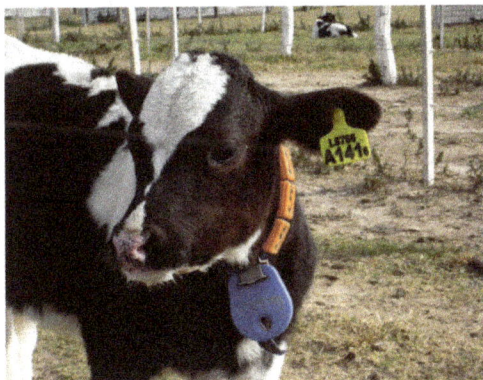

Calf identified with ear tag and transponder

History

Individual identification of animals by means of body markings has been practised for over 3,800 years, as stated in Code of Hammurabi. The first official identification systems are documented as far as the 18th century. In Uruguay for instance maintained at that time a register of hot brands.

Methods

Birds

- Leg rings
- Wing tags
- Microchip implants (parrots)
- Telemetry (falconry birds)

Sheep

- Freeze branding
- Branding (hot-iron)
- Collar
- Earmarking
- Ear tags (non-electronic)
- Ear tags (electronic)
- Semi-permanent paint

Pigs

- Collars (electronic and non-electronic)
- Earmarking
- Ear tags (non-electronic)
- Ear tags (electronic)
- Semi-permanent paint
- Tattoo

Horses

- Collars (non-electronic)
- Branding (hot-iron)

- Branding (freeze)
- Microchip implants
- Lip tattoo

Cattle

- Anklets
- Branding (freeze)
- Branding (hot-iron)
- Collars (electronic and non-electronic)
- Earmarking
- Ear tags (non-electronic)
- Ear tags (electronic)
- Rumen bolus (electronic)
- Cowbell

Dogs

- Collar
- Microchip implants
- Tattoo

Laboratory Mice

- Earmarking (notching or punching)
- Ear tags (nickel, copper or scannable 2D barcode tags)
- Microchip implants
- Hair dye
- Toe clipping
- Manual tattoos (tail, foot pad or ears)
- Automated tail tattoos

Fish

- Microchip implants
- Fin clipping
- Coded wire tag
- Passive integrated transponder
- Acoustic tag

Marine Mammals

- Transponders
- Adhesive tags

Invertebrates

- Adhesive tags
- Semi-permanent paint

Microchip Implant (Animal)

A microchip implant is an identifying integrated circuit placed under the skin of an animal. The chip, about the size of a large grain of rice, uses passive RFID (Radio Frequency Identification) technology, and is also known as a PIT tag (for Passive Integrated Transponder).

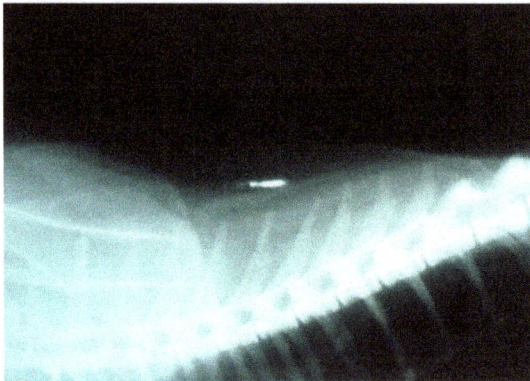

Microchip implant in a cat.

Externally attached microchips such as RFID ear tags are commonly used to identify farm and ranch animals other than horses. Some external microchips can be read with the same scanner used with implanted chips.

Uses and Benefits

Animal shelters, animal control officers and veterinarians routinely look for microchips to return lost pets quickly to their owners, avoiding expenses for housing, food, medical care, outplacing and euthanasia. Many shelters place chips in all outplaced animals.

Microchips are also used by kennels, breeders, brokers, trainers, registries, rescue groups, humane societies, clinics, farms, stables, animal clubs and associations, researchers, and pet stores.

Microchips can activate some pet doors programmed to recognize specific animals.

Some countries require microchips in imported animals to match vaccination records. Microchip tagging may also be required for CITES-regulated international trade in certain endangered animals: for example, Asian Arowana are tagged to limit import to captive-bred fish. Also, birds not banded who cross international borders as pets or for trade must be microchipped so that each bird is uniquely identifiable.

Usage

Microchips can be implanted by a veterinarian or at a shelter. After checking that the animal does not already have a chip, the vet or technician injects the chip with a syringe and records the chip's unique ID. No anesthetic is required. A test scan ensures correct operation.

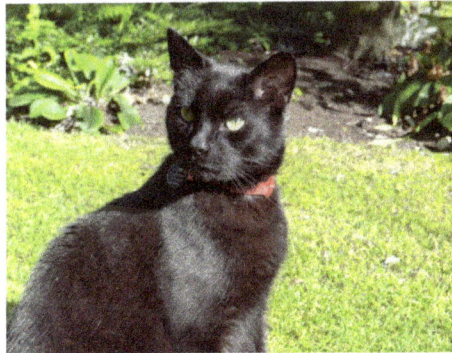

Information about the implant is often imprinted on a collar tag worn by a pet

An enrollment form is completed with chip ID, owner contact information, pet name and description, shelter and/or veterinarian contact information, and an alternate emergency contact designated by the pet owner. Some shelters and vets designate themselves as the primary contact to remain informed about possible problems with the animals they place. The form is sent to a registry, who may be the chip manufacturer, distributor or an independent entity such as a pet recovery service. Some countries have a single official national database. For a fee, the registry typically provides 24-hour, toll-free telephone service for the life of the pet. Some veterinarians leave registration to the owner, usually done online, but a chip without current contact information is essentially useless.

The owner receives a registration certificate with the chip ID and recovery service contact information. The information can also be imprinted on a collar tag worn by the animal. Like an automobile title, the certificate serves as proof of ownership and is transferred with the animal when it is sold or traded; an animal without a certificate could be stolen.

Authorities and shelters examine strays for chips, providing the recovery service with the ID number, description and location so they may notify the owner or contact. If the pet is wearing the collar tag, the finder does not need a chip reader to contact the registry. An owner can also report a missing pet to the recovery service, as vets look for chips in new animals and check with the recovery service to see if it has been reported lost or stolen.

Many veterinarians scan an animal's chip on every visit to verify correct operation. Some use the chip ID as their database index and print it on receipts, test results, vaccination certifications and other records.

Components of a Microchip

A microchip implant is a passive RFID device. Lacking an internal power source, it remains inert until it is powered by the scanner.

Most implants contain three elements: a 'chip' or integrated circuit; a coil inductor, possibly with a ferrite core; and a capacitor. The chip contains unique identification data and electronic circuits to encode that information. The coil acts as the secondary winding of a transformer, receiving power inductively coupled to it from the scanner. The coil and capacitor together form a resonant LC circuit tuned to the frequency of the scanner's oscillating magnetic field to produce power for the chip. The chip then transmits its data back through the coil to the scanner.

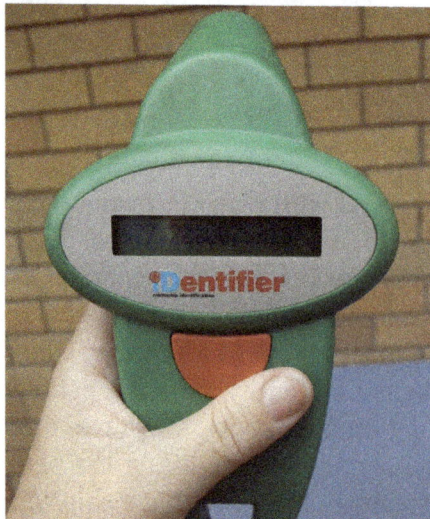

Example of an RFID scanner used with animal microchip implants.

These components are encased in biocompatible soda lime or borosilicate glass and hermetically sealed. Barring rare complications, dogs and cats are unaffected by them.

Implant Location

In dogs and cats, chips are usually inserted below the skin at the back of the neck between the shoulder blades on the dorsal midline. According to one reference, continental European pets get the implant in the left side of the neck. The chip can often be felt under the skin. Thin layers of connective tissue form around the implant and hold it in place.

Horses are microchipped on the left side of the neck, halfway between the poll and withers and approximately one inch below the midline of the mane, into the nuchal ligament.

Birds are implanted in their breast muscles. Proper restraint is necessary so the operation requires either two people (an avian veterinarian and a veterinary technician) or general anesthesia.

Implanted microchips can distort magnetic resonance imaging (MRIs), including those of the spinal cord.

Animal Species

Many animal species have been microchipped, including cockatiels and other parrots, horses, llamas, alpacas, goats, sheep, miniature pigs, rabbits, deer, ferrets, penguins, sharks, snakes, lizards, alligators, turtles, toads, frogs, rare fish, chimpanzees, mice, and prairie dogs—even whales and elephants. The U.S. Fish and Wildlife Service uses microchipping in its research of wild bison, black-footed ferrets, grizzly bears, elk, white-tailed deer, giant land tortoises and armadillos.

Horse microchipping

Worldwide Use

Microchips are not yet universal, but they are legally required in some jurisdictions such as the state of New South Wales, Australia and the United Kingdom (since 2016.04.06).

Some countries, such as Japan, require ISO-compliant microchips or a compatible reader on imported dogs and cats.

In New Zealand, all dogs first registered after 1 July 2006 must be microchipped. Farmers protested that farm dogs should be exempt, drawing a parallel to the Dog Tax War of 1898. Farm dogs were exempted from microchipping in an amendment to the legislation passed in June 2006. A National Animal Identification and Tracing scheme in New Zealand is currently being developed for tracking livestock.

In April 2012 Northern Ireland became the first part of the United Kingdom to require microchipping of individually licensed dogs. England will have mandatory microchipping of all dogs by 2016.

In Israel, microchips in dogs are mandatory.

Australia has a National Livestock Identification System.

The United States uses the National Animal Identification System for farm and ranch animals other than dogs and cats. In most species except horses, an external eartag is typically used in lieu of an implant microchip. Eartags with microchips or simply stamped with a visible number can be used. Both use ISO 15 digit microchip numbers with the U.S. country code of 840.

Cross-compatibility and Standards Issues

In most countries, pet ID chips adhere to an international standard to promote compatibility between chips and scanners. In the United States, however, three proprietary types of chips compete along with the international standard. Scanners distributed to United States shelters and veterinarians well into 2006 could each read at most three of the four types. Scanners with quad-read capability are now available and are increasingly considered required equipment. Older scanner models will be in use for some time, so United States pet owners must still choose between a chip with good coverage by existing scanners and one compatible with the international standard. The four types include:

- The ISO Conformant Full Duplex type has the greatest international acceptance. It is common in many countries including Europe (since the late 1990s) and Canada. It is one of two chip protocol types (along with the "Half Duplex" type sometimes used in farm and ranch animals) that conform to International Organization for Standardization standards ISO 11784 & 11785. To support international/multivendor application, the 3-digit country code can contain an assigned ISO country code or a manufacturer code from 900 to 998 plus its identifying serial number. In the United States, distribution of this type has been controversial. When 24PetWatch.com began distributing them in 2003 (and more famously Banfield Pet Hospitals in 2004) many shelter scanners couldn't read them. (Some still can't; asking local shelters about this is still a good idea.) At least one Banfield-chipped pet was inadvertently euthanized.

- The Trovan Unique type is another pet chip protocol type in use since 1990 in pets in the United States. Patent problems forced the withdrawal of Trovan's implanter device from United States distribution and they became uncommon in pets in the United States, although Trovan's original registry database "info-pet.biz" remained in operation. In early 2007, the American Kennel Club's chip registration service, AKC Companion Animal Recovery Corp, which had been the authorized registry for HomeAgain brand chips made by Destron/Digital Angel, began distributing Trovan chips with a different implanter. These chips are read by the Trovan, HomeAgain (Destron Fearing), and Bayer (Black Label) readers. Despite multiple offers from Trovan to AVID to license the technology to read the Trovan chips, AVID continues to distribute readers that do not read Trovan or the ISO compliant chips.

- A third type sometimes known as FECAVA or Destron is available under various brand names. These include, in the United States, "Avid Eurochip", the common current 24PetWatch chips, and the original (and still popular) style of HomeAgain chips. (HomeAgain and 24Petwatch can now supply the true ISO chip instead on request.) Chips of this type have 10 digit [hexadecimal] chip numbers. This "FECAVA" type is readable on a wide variety of scanners in the United States and has been less controversial, although its level of adherence to the ISO standards is sometimes exaggerated in some descriptions. The ISO standard has an annex (appendix) recommending that three older chip types be supported by scanners, including a 35-bit "FECAVA"/"Destron" type. The common Eurochip/HomeAgain chips don't agree perfectly with the annex description, although the differences are sometimes considered minor. But the ISO standard also makes it clear that only its 64-bit "full-duplex" and "half-duplex" types are "conformant"; even chips (e.g., the Trovan Unique) that match one of the Annex descriptions are not. More visibly, FECAVA cannot support the ISO standard's required country/manufacturer codes. They may be accepted by authorities in many countries where ISO-standard chips are the norm, but not by those requiring literal ISO conformance.

- Finally, there's the AVID brand Friendchip type, which is peculiar due to its encryption characteristics. Cryptographic features are not necessarily unwelcome; few pet rescuers or humane societies would object to a design that outputs an ID number "in the clear" for anyone to read, along with authentication features for detection of counterfeit chips, but the authentication in "Friendchips" has been found lacking and rather easy to spoof to the AVID scanner. Although no authentication encryption is involved, obfuscation requires proprietary information to convert transmitted chip data to its original label ID code. Well into 2006, scanners containing the proprietary decryption were provided to the United States market only by AVID and Destron/Digital Angel; Destron/Digital Angel put the decryption feature in some, but not all, of its scanners, possibly as early as 1996. (For years, its scanners distributed to shelters through HomeAgain usually had full decryption, while many sold to veterinarians would only state that an AVID chip had been found.) Well into 2006, both were resisting calls from consumers and welfare group officials to bring scanners to the United States shelter community combining AVID decryption capability with the ability to read ISO-compliant chips. Some complained that AVID itself had long marketed combination pet scanners compatible with all common pet chips except possibly Trovan outside the United States. By keeping them out of the United States, it could be considered partly culpable in the missed-ISO chips problem others blamed on Banfield. In 2006, the European manufacturer Datamars, a supplier of ISO chips used by Banfield and others, gained access to the decryption secrets and began supplying scanners with them to United States customers. This "Black Label" scanner was the first four-standard full-multi pet scanner in the United States market. Later in 2006, Digital Angel announced that it would supply a full-multi scanner in the United States. In 2008 AVID announced a "breakthrough" scanner, although as of October 2010 AVID's is still so uncommon that it's unclear whether it supports the Trovan chip. Trovan also acquired the decryption technology in 2006 or earlier, and now provides it in scanners distributed in the United States by AKC-CAR. (Some are quad-read, but others lack full ISO support.)

Numerous references in print state that the incompatibilities between different chip types are a matter of "frequency". One may find claims that early ISO adopters in the United States endangered their customers' pets by giving them ISO chips that work at a "different frequency" from the local shelter's scanner, or that the United States government considered forcing an incompatible frequency change. These claims were little challenged by manufacturers and distributors of ISO chips, although later evidence suggests the claims were disinformation. In fact, all chips operate at the scanner's frequency. Although ISO chips are optimized for 134.2 kHz, in practice they are readable at 125 kHz and the "125 kHz" chips are readable at 134.2 kHz. Confirmation comes from government filings that indicate the supposed "multi-frequency" scanners now commonly available are really single-frequency scanners operating at 125, 134.2 or 128 kHz. In particular, the United States HomeAgain scanner didn't change excitation frequency when ISO-read capability was added; it's still a single frequency, 125 kHz scanner.

Scanner Compatibility table for chip types used in pets		
	Expected results for chip type **(OK=Good read** **NR=No readDO=Detect Only with no number given)**	

Scanner to test	ISO Conformant Full Duplex chip	AVID Encrypted "FriendChip"	Original U.S. HomeAgain, AVID Eurochip, or FECAVA	"Trovan Unique" and current AKC CAR chips
Minimal ISO Conformant Scanner (also must read HALF Duplex chips common in livestock ear tags)	OK	NR	NR	NR
AVID Basic U.S. Scanner	NR	OK	NR	NR
AVID Deluxe U.S. Scanner	NR	OK	OK	NR
AVID Universal Scanner sold outside U.S.	OK	OK	OK	NR Assumed
AVID MiniTracker Pro Scanner announced August 2008	OK	OK	OK	NR according to some (Few have seen one.)
Various vintages of U.S. HomeAgain "Universal" Shelter Scanners by Destron/Digital Angel Corp.	NR, DO, or OK	OK	OK	Possibly all OK
Typical Destron/ Digital Angel Corp. U.S. Veterinarian's scanner pre-2007	NR	DO	OK	DO
Trovan LID-560-MULTI per mfr. specifications on Web	OK	OK	OK	OK
U.S. Trovan Pocket Scanner per AKC-CAR Web Site	DO	OK	OK	OK

U.S. Trovan ProS-can700 per AKC-CAR Web Site	OK	OK	OK	OK
Original 2006 Datamars Black Label Scanner	OK	OK	OK	OK but Reliability Questioned
Datamars Black Label Scanner "classypets" model	OK	NR or DO?	OK	OK but Reliability Questioned
Banfield-Distrib-uted 2004-2005 Vintage Datamars Scanners	OK	Possibly all DO	OK	Possibly all OK but Reliability Ques-tioned (Undocu-mented Feature)
Datamars Mini-max and Micro-max	OK	NR	NR	NR
Typical Home-made Scanner	OK	OK but extra step required (web-based decryption service)	OK	OK

(For users requiring Shelter-Grade certainty, this table is not a substitute for testing the scanner with a set of specimen chips. One study cites problems with certain Trovan chips on the Datamars Black Label scanner. In general, the study found none of the tested scanners to read all four standards without some deficiency. The study predates the most recent scanner models, however.)

Reported Adverse Reactions

RFID chips are used in animal research, and at least three studies conducted since the 1990s have reported tumors at the site of implantation in laboratory mice and rats. Noted veterinary associations responded with continued support for the procedure as reasonably safe for cats and dogs, pointing to rates of serious complications on the order of one in a million in the U.K., which has a system for tracking such adverse reactions and has chipped over 3.7 million pet dogs. A recent study found no safety concerns for microchipped animals with RFID chips undergoing MRI at one Tesla magnetic field strength. In 2011 a microchip-associated fibrosarcoma was reported found in the neck of a 9-year old, neutered-male cat. Histological examination was consistent with post-injection sarcoma, but all prior vaccinations occurred in the hindlegs.

Real-time Locating System

Real-time locating systems (RTLS) are used to automatically identify and track the location of objects or people in real time, usually within a building or other contained area.

Wireless RTLS tags are attached to objects or worn by people, and in most RTLS, fixed reference points receive wireless signals from tags to determine their location. Examples of real-time locating systems include tracking automobiles through an assembly line, locating pallets of merchandise in a warehouse, or finding medical equipment in a hospital.

The physical layer of RTLS technology is usually some form of radio frequency (RF) communication, but some systems use optical (usually infrared) or acoustic (usually ultrasound) technology instead of or in addition to RF. Tags and fixed reference points can be transmitters, receivers, or both, resulting in numerous possible technology combinations.

RTLS are a form of local positioning system, and do not usually refer to GPS or to mobile phone tracking. Location information usually does not include speed, direction, or spatial orientation.

Origin

The term RTLS was created (circa 1998) at the ID EXPO trade show by Tim Harrington (WhereNet), Jay Werb, (PinPoint), and Bert Moore, (Automatic Identification Manufacturers, Inc.(AIM)). It was created to describe and differentiate an emerging technology that not only provided the automatic identification capabilities of active RFID tags, but also added the ability to view the location on a computer screen. It was at this show that the first examples of a commercial radio based RTLS system were shown by PinPoint and WhereNet. Although this capability had been utilized previously by military and government agencies, the technology had been too expensive for commercial purposes. In the early 1990s, the first commercial RTLS were installed at three healthcare facilities in the United States, and were based on the transmission and decoding of infrared light signals from actively transmitting tags. Since then, new technology has emerged that also enables RTLS to be applied to passive tag applications.

Locating Concepts

RTLS are generally used in indoor and/or confined areas, such as buildings, and do not provide global coverage like GPS. RTLS tags are affixed to mobile items to be tracked or managed. RTLS reference points, which can be either transmitters or receivers, are spaced throughout a building (or similar area of interest) to provide the desired tag coverage. In most cases, the more RTLS reference points that are installed, the better the location accuracy, until the technology limitations are reached.

A number of disparate system designs are all referred to as "real-time locating systems", but there are two primary system design elements:

Locating at Choke Points

The simplest form of choke point locating is where short range ID signals from a moving tag are received by a single fixed reader in a sensory network, thus indicating the location coincidence of reader and tag. Alternately, a choke point identifier can be received by the moving tag, and then relayed, usually via a second wireless channel, to a location processor. Accuracy is usually defined by the sphere spanned with the reach of the choke point transmitter or receiver. The use of directional antennas, or technologies such as infrared or ultrasound that are blocked by room partitions, can support choke points of various geometries.

Locating in Relative Coordinates

ID signals from a tag is received by a multiplicity of readers in a sensory network, and a position is estimated using one or more locating algorithms, such as trilateration, multilateration, or triangulation. Equivalently, ID signals from several RTLS reference points can be received by a tag, and relayed back to a location processor. Localization with multiple reference points requires that distances between reference points in the sensory network be known in order to precisely locate a tag, and the determination of distances is called ranging.

Another way to calculate relative location is if mobile tags communicate directly with each other, then relay this information to a location processor.

Location Accuracy

RF trilateration uses estimated ranges from multiple receivers to estimate the location of a tag. RF triangulation uses the angles at which the RF signals arrive at multiple receivers to estimate the location of a tag. Many obstructions, such as walls or furniture, can distort the estimated range and angle readings leading to varied qualities of location estimate. Estimation-based locating is often measured in accuracy for a given distance, such as 90% accurate for 10 meter range.

Systems that use locating technologies that do not go through walls, such as infrared or ultrasound, tend to be more accurate in an indoor environment because only tags and receivers that have line of sight (or near line of sight) can communicate.

Applications

RTLS can be used numerous logistical or operational areas such as:

- locate and manage assets within a facility, such as finding a misplaced tool cart in a warehouse or medical equipment

- notification of new locations, such as an alert if a tool cart improperly has left the facility

- to combine identity of multiple items placed in a single location, such as on a pallet

- to locate customers, for example in a restaurant, for delivery of food or service

- to maintain proper staffing levels of operational areas, such as ensuring guards are in the proper locations in a correctional facility

- to quickly and automatically account for all staff after or during an emergency evacuation

- to automatically track and time stamp the progress of people or assets through a process, such as following a patient's emergency room wait time, time spent in the operating room, and total time until discharge. Such a system can be used for process improvement

- clinical-grade locating to support acute care capacity management

Privacy Concerns

RTLS may be seen as a threat to privacy when used to determine the location of people. The newly declared human right of informational self-determination de:Informationelle Selbstbestimmung gives the right to prevent one's identity and personal data from being disclosed to others, and also covers disclosure of locality, though this does not generally apply to the workplace.

Several prominent labor unions have come out against the use of RTLS systems to track workers calling them "the beginning of Big Brother" and "an invasion of privacy". However, this loss of privacy may be outweighed by other benefits to staff. For example, Toronto General Hospital is looking at RTLS to reduce quarantine times after an infectious disease outbreak. After a recent SARS outbreak, 1% of all staff were quarantined, and more accurate data regarding who had been exposed to the virus could have reduced the need for quarantines.

Types of Technologies Used

There is a wide variety of systems concepts and designs to provide real-time locating.

- Active radio frequency identification (Active RFID)

- Active radio frequency identification - infrared hybrid (Active RFID-IR)

- Infrared (IR)

- Optical locating

- Low-frequency signpost identification

- Semi-active radio frequency identification (semi-active RFID)

- Passive RFID RTLS locating via Steerable Phased Array Antennae

- Radio beacon,

- Ultrasound Identification (US-ID)

- Ultrasonic ranging (US-RTLS)

- Ultra-wideband (UWB)

- Wide-over-narrow band

- Wireless Local Area Network (WLAN, Wi-Fi)

- Bluetooth,

- Clustering in noisy ambience,

- Bivalent systems

A general model for selection of the best solution for a locating problem has been constructed at the Radboud University of Nijmegen. Many of these references do not comply with the definitions given in international standardization with ISO/IEC 19762-5 and ISO/IEC 24730-1. However, some aspects of real-time performance are served and aspects of locating are addressed in context of absolute coordinates.

Ranging and Angulating

Depending on the physical technology used, at least one and often some combination of ranging and/or angulating methods are used to determine location:

- Angle of arrival (AoA)

- Line-of-sight (LoS)

- Time of arrival (ToA)

- Multilateration (Time difference of arrival) (TDoA)

- Time-of-flight (ToF)

- Two-way ranging (TWR) according to Nanotron's patents

- Symmetrical Double Sided – Two Way Ranging (SDS-TWR)

- Near-field electromagnetic ranging (NFER)

Errors and Accuracy

Real-time locating is affected by a variety of errors. Many of the major reasons relate to the physics of the locating system, and may not be reduced by improving the technical equipment.

None or No Direct Response

Many RTLS systems require direct and clear line of sight visibility. For those systems, where there is no visibility from mobile tags to fixed nodes there will be no result or a non valid result from locating engine. This applies to satellite locating as well as other RTLS systems such as angle of arrival and time of arrival. Fingerprinting is a way to overcome the visibility issue: If the locations in the tracking area contain distinct measurement fingerprints, line of sight is not necessarily needed. For example, if each location contains a unique combination of signal strength readings from transmitters, the location system will function properly. This is true, for example, with some Wi-Fi based RTLS solutions. However, having distinct signal strength fingerprints in each location typically requires a fairly high saturation of transmitters.

False Location

The measured location may appear entirely faulty. This is a generally result of simple operational models to compensate for the plurality of error sources. It proves impossible to serve proper location after ignoring the errors.

Locating Backlog

Real time is no registered branding and has no inherent quality. A variety of offers sails under this term. As motion causes location changes, inevitably the latency time to compute a new location may be dominant with regard to motion. Either an RTLS system that requires waiting for new results is not worth the money or the operational concept that asks for faster location updates does not comply with the chosen systems approach.

Temporary Location Error

Location will never be reported exactly, as the term real-time and the term precision directly contradict in aspects of measurement theory as well as the term precision and the term cost contradict in aspects of economy. That is no exclusion of precision, but the limitations with higher speed are inevitable.

Steady Location Error

Recognizing a reported location steadily apart from physical presence generally indicates the problem of insufficient over-determination and missing of visibility along at least one link from resident anchors to mobile transponders. Such effect is caused also by insufficient concepts to compensate for calibration needs.

Location Jitter

Noise from various sources has an erratic influence on stability of results. The aim to provide a steady appearance increases the latency contradicting to real time requirements.

Location Jump

As objects containing mass have limitations to jump, such effects are mostly beyond physical reality. Jumps of reported location not visible with the object itself generally indicate improper modeling with the location engine. Such effect is caused by changing dominance of various secondary responses.

Location Creep

Location of residing objects gets reported moving, as soon as the measures taken are biased by secondary path reflections with increasing weight over time. Such effect is caused by simple averaging and the effect indicates insufficient discrimination of first echoes.

Standards

ISO/IEC

The basic issues of RTLS are standardized by the International Organization for Standardization and the International Electrotechnical Commission, under the ISO/IEC 24730 series. In this series of standards, the basic standard ISO/IEC 24730-1 identifies the terms describing a form of RTLS used by a set of vendors, but does not encompass the full scope of RTLS technology.

Currently several standards are published:

- ISO/IEC 19762-5:2008 Information technology — Automatic identification and data capture (AIDC) techniques — Harmonized vocabulary—Part 5: Locating systems

- ISO/IEC 24730-1:2014 Information technology — Real-time locating systems (RTLS) — Part 1: Application programming interface (API)

- ISO/IEC 24730-2:2012 Information technology — Real time locating systems (RTLS) — Part 2: Direct Sequence Spread Spectrum (DSSS) 2,4 GHz air interface protocol

- ISO/IEC 24730-5:2010 Information technology — Real-time locating systems (RTLS) — Part 5: Chirp spread spectrum (CSS) at 2,4 GHz air interface

- ISO/IEC 24730-21:2012 Information technology — Real time locating systems (RTLS) — Part 21: Direct Sequence Spread Spectrum (DSSS) 2,4 GHz air interface protocol: Transmitters operating with a single spread code and employing a DBPSK data encoding and BPSK spreading scheme

- ISO/IEC 24730-22:2012 Information technology — Real time locating systems (RTLS) — Part 22: Direct Sequence Spread Spectrum (DSSS) 2,4 GHz air interface protocol: Transmitters operating with multiple spread codes and employing a QPSK data encoding and Walsh offset QPSK (WOQPSK) spreading scheme

- ISO/IEC 24730-61:2013 Information technology — Real time locating systems

(RTLS) — Part 61: Low rate pulse repetition frequency Ultra Wide Band (UWB) air interface

- ISO/IEC 24730-62:2013 Information technology — Real time locating systems (RTLS) — Part 62: High rate pulse repetition frequency Ultra Wide Band (UWB) air interface

These standards do not stipulate any special method of computing locations, nor the method of measuring locations. This may be defined in specifications for trilateration, triangulation or any hybrid approaches to trigonometric computing for planar or spherical models of a terrestrial area.

Incits

- INCITS 371.1:2003, Information Technology - Real Time Locating Systems (RTLS) - Part 1: 2.4 GHz Air Interface Protocol

- INCITS 371.2:2003, Information Technology - Real Time Locating Systems (RTLS) - Part 2: 433-MHz Air Interface Protocol

- INCITS 371.3:2003, Information Technology - Real Time Locating Systems (RTLS) - Part 3: Application Programming Interface

Limitations and Further Discussion

In RTLS application in the Healthcare industry, various studies were issued discussing the limitations of the currently adopted RTLS. Currently used technologies RFID, Wi-fi, UWB, all RFID based are hazardous in the sense of interference with sensitive equipment. A study carried out by Dr Erik Jan van Lieshout of the Academic Medical Centre of the University of Amsterdam published in 'JAMA' (Journal of the American Medical Equipment) claimed "RFID and UWB could shut down equipment patients rely on" as "RFID caused interference in 34 of the 123 tests they performed". The first Bluetooth RTLS provider in the medical industry is supporting this in their article: "The fact that RFID cannot be used near sensitive equipment should in itself be a red flag to the medical industry". The RFID Journal responded to this study not negating it rather explaining real-case solution: "The Purdue study showed no effect when ultrahigh-frequency (UHF) systems were kept at a reasonable distance from medical equipment. So placing readers in utility rooms, near elevators and above doors between hospital wings or departments to track assets is not a problem". However the case of 'keeping at a reasonable distance' might be still an open question for the RTLS technology adopters and providers in medical facilities.

In many applications it is very difficult and at the same time important to make a proper choice among various communication technologies (e.g., RFID, WiFi, etc.) which RTLS may include. Wrong design decision made at early stages can lead to catastrophic results fore the system and a significant loss of money for fixing and redesign. To solve this problem a special metodology for RTLS design space exploration was developed. It consists of such steps as modelling, requirements specification and verification into a single efficient process.

Electronic Article Surveillance

Electronic article surveillance (EAS) is a technological method for preventing shoplifting from retail stores, pilferage of books from libraries or removal of properties from office buildings. Special tags are fixed to merchandise or books. These tags are removed or deactivated by the clerks when the item is properly bought or checked out. At the exits of the store, a detection system sounds an alarm or otherwise alerts the staff when it senses active tags. Some stores also have detection systems at the entrance to the restrooms that sound an alarm if someone tries to take unpaid merchandise with them into the restroom. For high-value goods that are to be manipulated by the patrons, wired alarm clips called spider wrap may be used instead of tags.

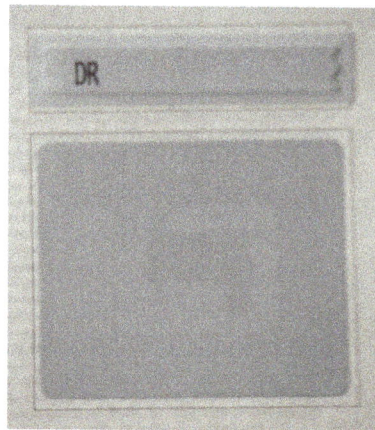

Electronic article surveillance tags: acousto-magnetic (top) and RF (bottom).

Types

There are several major types of electronic article surveillance systems:

- Electro-Magnetic, also known as magneto-harmonic

- Acousto-magnetic, also known as magnetostrictive

- Radio Frequency (8,2 MHz)

- Microwave

- Video surveillance systems (to some extent)

Electro-magnetic Systems

These tags are made of a strip of amorphous metal (metglas) which has a very low magnetic saturation value. Except for permanent tags, this strip is also lined with a strip of ferromagnetic material with a moderate coercive field (magnetic "hardness"). Detection is achieved by sensing harmonics and sum or difference signals generated by the non-linear magnetic response of the material under a mixture of low-frequency (in the 10 Hz to 1000 Hz range) magnetic fields.

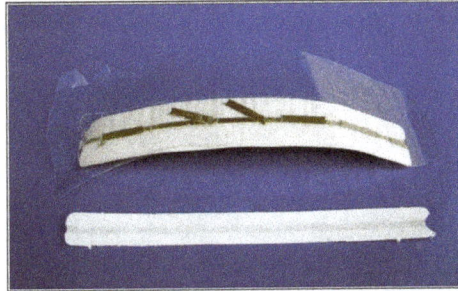

EM tag

When the ferromagnetic material is magnetized, it biases the amorphous metal strip into saturation, where it no longer produces harmonics. Deactivation of these tags is therefore done with magnetization. Activation requires demagnetization.

This system is suitable for items in libraries since the tags can be deactivated when items are borrowed and re-activated upon return. It is also suitable for merchandise in retail stores, due to the small size and very low cost of the tags (Tattle-Tape).

Electro-Magnetic systems are no longer common in Food or Apparel retail environments, most having been replaced by Radio Frequency, Acousto-Magnetic or even RFID platforms. The technology is still popular in Library environments where the need to deactivate and reactivate tags is a necessity, however here too RFID platforms have been gaining ground in the last 10 years due to their greater range of use.

Acousto-magnetic Systems

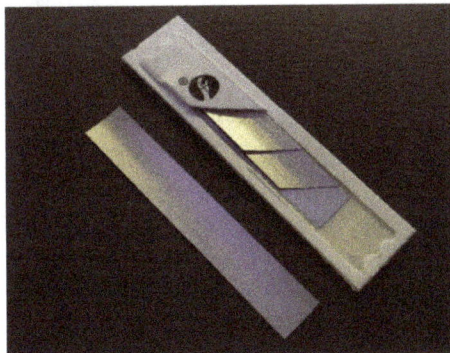

A cutaway image of an acousto-magnetic tag.

These are similar to magnetic tags in that they are made of two strips, a strip of magnetostrictive, ferromagnetic amorphous metal and a strip of a magnetically semi-hard metallic strip, which is used as a biasing magnet (to increase signal strength) and to allow deactivation. These strips are not bound together but free to oscillate mechanically.

Amorphous metals are used in such systems due to their good magnetoelastic coupling, which implies that they can efficiently convert magnetic energy into mechanical vibrations.

The detectors for such tags emit periodic tonal bursts at about 58 kHz, the same as the resonance frequency of the amorphous strips. This causes the strip to vibrate longitudinally by magnetostriction, and it continues to oscillate after the burst is over. The vibration causes a change in magnetization in the amorphous strip, which induces an AC voltage in the receiver antenna. If this signal meets the required parameters (correct frequency, repetition, etc.), the alarm is activated.

When the semi-hard magnet is magnetized, the tag is activated. The magnetized strip makes the amorphous strip respond much more strongly to the detectors, because the DC magnetic field given off by the strip offsets the magnetic anisotropy within the amorphous metal. The tag can also be deactivated by demagnetizing the strip, making the response small enough so that it will not be detected by the detectors.

AM tags are three dimensional plastic tags, much thicker than electro-magnetic strips and are thus seldom used for books. However, they are relatively inexpensive and have better detection rates (fewer false positives and false negatives) than magnetic tags.

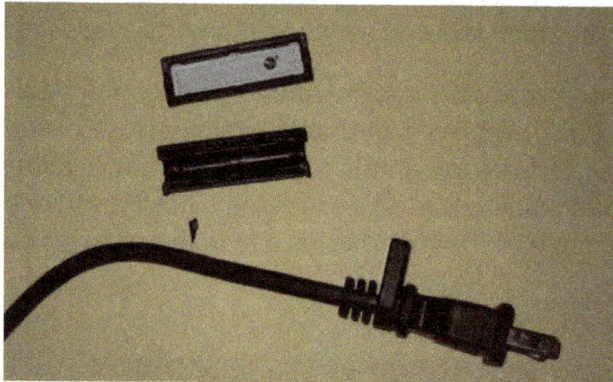

Emtag-removed

Called Emtag by B&G International, this type tag is often attached to the inside of a plastic surround permanently attached to the power cords of hand tools and equipment.

Radio Frequency Systems

These tags are essentially an LC tank circuit that has a resonance peak anywhere from 1.75 MHz to 9.5 MHz. The standard frequency for retail use is 8.2 MHz. Sensing is achieved by sweeping around the resonant frequency and detecting the dip.

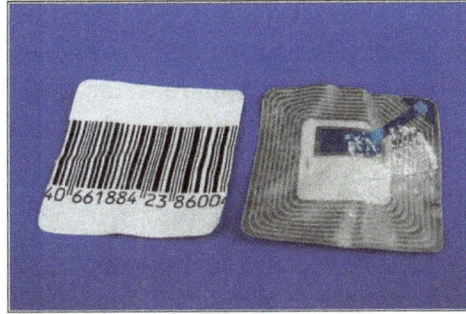

RF label

Deactivation for 8.2 MHz label tags is typically achieved using a deactivation pad. In the absence of such a device labels can be rendered inactive by punching a hole, or by covering the circuit with a metallic label, a "detuner". The deactivation pad functions by partially destroying the capacitor, though this sounds violent, in reality both the process and the result are unnoticeable to the naked eye, the deactivator causes a micro short circuit in the label. This is done by submitting the tag to a strong electromagnetic field at the resonant frequency, which induces voltages exceeding the capacitor's breakdown voltage.

Series 304 RF EAS label

In terms of deactivation Radio Frequency is the most efficient of the 3 technologies (RF, EM, AM - there are no microwave labels) given that the reliable "remote" deactivation distance can be to 30 cm. It also benefits the user in terms of running costs since the RF deactivator only activates to send a pulse when a circuit is present. Both EM and AM deactivation units are on all the time and consume considerably more electricity. The reliability of "remote" deactivation (i.e. non contact or non proximity deactivation) capability makes for a fast and efficient throughput at the checkout.

Efficiency is an important factor when choosing an overall EAS solution given that time lost attempting to deactivate labels can be an important drag of cashier productivity as well as customer satisfaction if unwanted alarms are caused by tags that have not been effectively deactivated at the point of sale.

Deactivation of RF labels is also dependent on the size of the label and the power of the deactivation pad (the larger the label, the greater the field it generates for deactivation to take place. For this reason very small labels can cause issues for consistent deactivation). It is common to find RF deactivation built into barcode flat and vertical scanners at the POS in food retail especially in Europe and Asia where RF EAS technology has been the standard for nearly a decade. In Apparel retail deactivation usually takes the form of flat pads of approx. 30x30 cm in retail apparel environments.

Microwave Systems

These permanent tags are made of a non-linear element (a diode) coupled to one microwave and one electrostatic antenna. At the exit, one antenna emits a low-frequency (about 100 kHz) field, and another one emits a microwave field. The tag acts as a mixer re-emitting a combination of signals from both fields. This modulated signal triggers the alarm. These tags are permanent and somewhat costly. They are mostly used in clothing stores and have practically been withdrawn from use.

Source Tagging

Source tagging is the application of EAS security tags at the source, the supplier or manufacturer, instead of at the retail side of the chain. For the retailer, source tagging eliminates the labor expense needed to apply the EAS tags themselves, and reduces the time between receipt of merchandise and when the merchandise is ready for sale. For the supplier, the main benefit is the preservation of the retail packaging aesthetics by easing the application of security tags within product packaging. Source tagging allows the EAS tags to be concealed and more difficult to remove.

The high-speed application of EAS labels, suited for commercial packaging processes, was perfected via modifications to standard pressure-sensitive label applicators and was developed and introduced by Craig Patterson, initially for Hewlett Packard print cartridges. Today, consumer goods are source tagged at high speeds with the EAS label incorporated into the packaging or the product itself.

The most common source tags are AM strips and 8.2 MHz radio frequency labels. Most manufacturers use both when source tagging in the USA. In Europe there is little demand for AM tagging given that the Food and Department Store environments are dominated by RF technology.

One significant problem from source tagging is something called "tag pollution" caused when non-deactivated tags carried around by customers cause unwanted alarms, decreasing the effectiveness and integrity of the EAS system. The problem is that no store has both systems. Therefore, if a store actually has an anti-shoplifting system to deactivate a label they will only deactivate one of the two. This is often the reason why people trigger an alarm entering a store, which can cause great frustration for both customers and staff. The problem is most evident in shopping malls where customers wander between stores. Retailers who use types of loss-prevention systems other than AM or 8.2 MHz radio frequency systems will not be as affected by "tag pollution".

Discussion

Occasional Vs. Professional Shoplifters

EAS systems can provide a solid deterrent against casual theft. The occasional shoplifter, not being familiar with these systems and their mode of operation, will either get caught by them, or preferably, will be dissuaded from attempting any theft in the first place.

Informed shoplifters are conscious of how tags can be removed or deactivated. A common method of defeating RF tags is the use of so-called booster bags. These are typically large paper bags as used by high street retailers that have been lined with multiple layers of aluminium foil to effectively shield the RF label from detection, much like a faraday cage. A simile would be the loss of signal that a cell phone suffers inside an elevator, the electro-magnetic, or radio, waves are effectively blocked, reducing the ability to send or receive information.

However, they may miss some tags or be unable to remove or deactivate all of them, especially if concealed or integrated tags are used. As a service to retailers, many manufacturers integrate security tags in the packaging of their products, or even inside the product itself, though this is rare and not especially desirable either for the retailer or the manufacturer. The practical totality of EAS labels are discarded with the product packaging, this is of particular application in everyday items that consumers might carry on their person, nobody wants the inconvenience of potentially live reactivated EAS tags on their person when walking in and out of retail stores.

Hard Tags, typically used for clothing or Ink Tags, known as benefit denial tags, may reduce the rate of tag manipulation. Also, deactivating or detaching tags may be spotted by the shop staff.

Shoplifting tools are illegal in many jurisdictions, and can, in any case, serve as evidence against the perpetrators. Hence, informed shoplifters, although they decrease their risk of being caught by the EAS, expose themselves to much greater judicial risks if they get caught with tools, booster bags, or while trying to remove tags, as this characterizes intent to steal.

The possession of shoplifting tools (e.g. lined bags or wire cutters to cut bottle tags) can lead to the suspect being arrested for suspicion of theft or "Going equipped for stealing, etc." within the UK judicial system.

In summary, while even the least expensive EAS systems will catch most occasional shoplifters, a broader range of measures are still required for an effective response that can protect profits without impeding sales.

Installation Costs

A single EAS detector, suitable for a small shop, is accessible to all retail stores, and should form a part of any coherent loss or profit protection system. The cost of antennas has dropped considerably over time, a simple 2 antenna installation can frequently be had for less than 2k€.

Disposable tags cost a matter of cents and may have been embedded during manufacture. More sophisticated systems are available, which are more difficult to circumvent. These solutions tend to be product category specific as in the case of high value added electronics and consumables, in consequence they are more expensive. Examples are "Safers", transparent secure boxes that completely enclose the article to be protected, Spiders that wrap around packaging and Electronic Merchandise Security Systems that allow phones and tablets to be used securely in the store before purchase. All of these require specific detachers or electronic keys at the Point Of Sale. They have the advantage of being reusable and being a strong visual deterrent to potential theft.

Tag Orientation

Except for microwave, the detection rate for all these tags depends on their orientation relative to the detection loops. For a pair of planar loops forming a Helmholtz coil, magnetic field lines will be approximately parallel in their center. Orienting the tag so that no magnetic flux from the coils crosses them will prevent detection, as the tag won't be coupled to the coils. This shortcoming, documented in the first EAS patents, can be solved by using multiple coils or by placing them in another arrangement such as a figure-of-eight. Sensitivity will still be orientation-dependent but detection will be possible at all orientations.

Detaching

A detacher is used to remove re-usable hard tags. The type of detacher used will depend on the type of tag. There are a variety of detachers available, with the majority using powerful magnets. Any store that uses an anti-shoplifting system and has a detacher should take care to keep it secured such that it cannot be removed. Some detachers actually have security tags inside them, to alert store personnel of them being removed from (or being brought into) the store. With greater frequency stores have metal detectors at the entrance that can warn against the presence of booster bags or illegal detachers.

Electro-magnetic Activation and Deactivation

Deactivation of magnetic tags is achieved by straightforward magnetization using a strong magnet. Magneto-acoustic tags require demagnetization. However, sticking a powerful magnet on them will bias disposable magnetic tags and prevent resonance in magneto-acoustic tags. Similarly, sticking a piece of metal, such as a large coin on a disposable radio-frequency tag will shield it. Non-disposable tags require stronger magnets or pieces of metal to disable or shield since the strips are inside the casing and thus further away. Deactivation of some EAS tags can trigger signals sent to the cashier alerting them.

Shielding

Most systems can be circumvented by placing the tagged goods in a bag lined with aluminum foil. The booster bag will act as a Faraday cage, shielding the tags from the antennas. Although some vendors claim that their acousto-magnetic systems cannot be defeated by bags shielded with aluminum foil, a sufficient amount of shielding (in the order of 30 layers of standard 20 μm foil) will defeat all standard systems.

Although the amount of shielding required depends on the system, its sensitivity, and the distance and orientation of the tags relative to its antennas, total enclosure of tags is not strictly necessary. Indeed, some shoplifters use clothes lined with aluminum foil. Low-frequency magnetic systems will require more shielding than radio-frequency systems due to their use of near-field magnetic coupling. Magnetic shielding, with steel or mu-metal, would be more effective, but also cumbersome and expensive.

The shielding technique is well-known amongst shoplifters and store owners. Some countries have specific laws against it. In any case, possession of such a bag demonstrates a prior-intent to commit a crime, which in many jurisdictions raises shoplifting from misdemeanor to felony status, as they are considered a "burglary tool."

To deter the use of booster bags, some stores have add-on metal detector systems which sense metallic surfaces.

Jamming

Like most systems that rely on transmission of electromagnetic signals through a hostile medium, EAS sensors can be rendered inoperative by jamming. As the signals from tags are very low-power (their cross-section is small, and the exits are wide), jamming requires little power. Evidently, shoplifters will not feel the need to follow radio transmission regulations; hence crude, easy-to-build transmitters will be adequate for them. However, due to their high frequency of operation, building a jammer can be difficult for microwave circuits; these systems are therefore less likely to be jammed. Although jamming is easy to perform, it is also easy to detect. A simple firmware upgrade should be adequate for modern DSP-based EAS systems to detect jamming. Nevertheless, the vast majority of EAS systems do not currently detect jamming.

Interference and Health Issues

All electronic article surveillance systems emit electromagnetic energy and thus can interfere with electronics.

Magneto-harmonic systems need to bring the tags to magnetic saturation and thus create magnetic fields strong enough to be felt through a small magnet. They routinely interfere with CRT displays. Demagnetization-remagnetization units also create intense fields.

Acousto-magnetic systems use less power but their signals are pulsed in the 100 Hz range.

Radio-frequency systems tend to be the least interfering because of their lower power and operating frequency in the MHz range, which makes it easy to shield against.

A March 2007 study by the Mayo Clinic in Rochester, Minnesota reported instances where acousto-magnetic EAS systems located at the front of retail stores caused a pacemaker to fail and a defibrillator to trigger, shocking the persons they were implanted in.

There are also concerns that some installations are intentionally reconfigured to exceed the rated specifications of the manufacturer, thereby exceeding tested and certified magnetic field levels.

Patents

Radio-frequency systems have the most manufacturers because they are not covered by patents, use well-known technology derived from radio communications, use little power, and can be manufactured without expensive metal alloys. There are exceptions however for example; a patented die-cut circuit manufacturing process is held by Miyake Inc. of Japan. Also, patents have recently been appearing for various combination tag designs for integration of both RF and RFID. The last known patent is the "Global Guard" by Argos Global, an EAS system that integrates video surveillance hidden inside pedestals, getting a close up of potential thieves.

Until recently, acousto-magnetic systems were covered by now-expired patents held by Sensormatic. WG Security Products, Inc. won a court battle against Sensormatic clarifying that WG Security Products acousto-magnetic systems did not infringe any Sensormatic patents. Disposable acousto-magnetic tags require special metal alloys; non-disposable ones require more expensive ferrite cores.

Contactless Payment

Contactless payment systems are credit cards and debit cards, key fobs, smart cards or other devices, including smartphones and other mobile devices, that use radio-frequency identification (RFID) or near field communication (NFC) for making secure payments. The embedded chip and antenna enable consumers to wave their card, fob, or handheld device over a reader at the point of sale terminal.

EMV contactless symbol used on compatible payment terminals

Some suppliers claim that transactions can be almost twice as fast as a conventional cash, credit, or debit card purchase. Because no signature or PIN verification is typically required, contactless purchases are typically limited. Those unauthorised may still take advantage of contactless payment systems as no identification occurs before payment except for certain devices, such as when using mobile payments. However, owners can block transactions, and that may provide a relatively short time frame, if any, for fraudulent activities to occur.

A contactless enabled American Express charge card issued in the UK

Research indicates that consumers are likely to spend more money using their cards due to the ease of small transactions. MasterCard Canada says it has seen "about 25 percent" higher spending by users of its Mastercard Contactless-brand RFID credit cards.

EMV is a common standard used by major credit card and smartphone companies for use in general commerce. Contactless smart cards that function as stored-value cards are becoming popular for use as transit system farecards, such as the Oyster card or RioCard. These can often store non-currency value (such as monthly passes) in additional to fare value purchased with cash or electronic payment.

History

Mobil was one of the most notable early adopters of this technology, and offered their "Speedpass" contactless payment system for participating Mobil gas stations as early as 1997. Although Mobil has since merged with Exxon, the service is still offered at many of ExxonMobil's stations. Freedompay also had early wins in the contactless space with Bank of America and McDonald's

McDonald's, KFC, Burger King, Boots, Eat, Heron Foods, Pret a Manger, Stagecoach Group, Subway, AMT Coffee, Tesco, Asda and Lidl are among the retailers offering contactless payments to their customers in the UK. In March 2008, EAT. became the first restaurant chain to adopt contactless.

Major financial entities now offering contactless payment systems include MasterCard, Citibank, JPMorgan Chase, American Express, KeyBank, Barclays, Barclaycard, HSBC,

Lloyds Banking Group, Freedompay, The Co-operative Bank, Nationwide Building Society and The Royal Bank of Scotland Group. Visa payWave, American Express Expresspay, and MasterCard Contactless are examples of contactless credit cards which have become widespread in U.S. and UK. The UK (and the rest of the world) version of the contactless applications differ from the U.S. one. The UK version has the capability of transacting offline, based on the limit stored in the application.

The first contactless cards in the UK were issued by Barclaycard in 2007. As of December 2014, there are approximately 58 million contactless-enabled cards in circulation in the UK and over 147,000 terminals in use though this is growing in numbers and percentages of adoption.

Telecom operators are starting to get involved in contactless payments via the use of near field communication phones. Belgacom's Pingping - Belgium, for example, has a stored value account and via a partnership with Alcatel-Lucent's touchatag provides contactless payment functionalities. In January 2010, Barclaycard partnered with mobile phone firm Orange, to launch a contactless credit card in the UK. Orange and Barclaycard also announced in 2009 that they will be launching a mobile phone with contactless technology.

In October 2011, the first mobile phones with MasterCard PayPass and/or Visa payWave certification appeared. A PayPass or payWave account can be assigned to the embedded secure element and/or SIM card within the phones. Android Pay is an application for devices running Google's Android OS, which allows users to make purchases using NFC, which initially required a physical secure element but this was replaced by host card emulation which was introduced in Android 4.4 (KitKat). Softcard (formerly known as Isis mobile wallet), Cityzi and Quick Tap wallets for example, use a secure SIM card to store encrypted personal information. Contactless payments with enabled mobile phones still occur on a small scale, but every month an increasing number of mobile phones are certified.

In February 2014, MasterCard announced that it would partner with Weve, which is a joint venture between EE, Telefónica UK, and Vodafone UK, to focus on mobile payments. The partnership will promote the development of "contactless mobile payment systems" by creating a universal platform in Europe for it.

In September of 2014, Transport for London's Tube began accepting contactless payment. The number of completed contactless journeys has now exceeded 300m. On Friday 18 December, the busiest single day in 2015, a record 1.24m journeys were completed by over 500k unique contactless cards.

In 2016 Erste Group launched an NFC only debit card implemented as a sticker in Austria. It can be used at any NFC supporting terminal for transactions of unlimited amount however for transactions over the floor limit of 25 EUR a PIN is required to confirm the transaction.

Security

As with all payment devices, contactless cards have a number of security features. Depending on the economic space, there may be a payment limit on single transactions, and some contactless cards can only be used a certain number of times before customers are asked for their PIN. Contactless debit and credit transactions use the same chip and PIN network as older cards are protected by the same fraud guarantees. Where PIN is supported (not in the United States), the contactless part of the card remains non-functional until a standard chip and PIN transaction has been executed. This provides some verification that the card was delivered to the actual cardholder. To reduce the likelihood of identity theft, a contactless card does not contain the card the cardholder's actual name; instead, a placeholder name is used.

Under fraud guarantee standards U.S. banks are liable for any fraudulent transactions charged to the contactless cards.

Floor Limit

Because no signature or PIN verification is typically required, contactless purchases are typically limited to a set amount per transaction, known as a "Floor Limit.".

Economic space	Limit	Comment
Australia	No limit	For transactions over A$100 a PIN is required.
Austria	No limit	For transactions over €25 a PIN is required. Additionally only three transactions can be made without a PIN.
Canada	No limit	Limits are completely at the discretion of the merchant's acquiring bank and the consumer's bank. There is no law limiting the amounts.
Chile	$12.000 CLP	
China	CNY ¥300	Unionpay QuickPass
Croatia	No limit	For transactions over 100HRK PIN or signature are needed.
Czech Republic	No limit	For transactions over 500 CZK PIN is needed. For every 3 consecutive contactless transactions PIN is needed.
Denmark	No limit	For transactions over 200DKK PIN is needed. Sometimes PIN is needed anyway to ensure the card is used by its owner.
Eurozone	€25	In general
Finland	€25	
France	€20	
Hong Kong	HKD $500	Some merchants can accept contactless payment if the transaction amount is under HKD $1000
Hungary	No limit	For transactions over 5000 HUF PIN is needed. For every 3 consecutive contactless transactions PIN is needed.
India	Rs. 2000	Above Rs. 2000 Contact chip transaction needs to be done.

Ireland	€30	Previously €15 until 1 October 2015.
Macedonia, Republic of	750 MKD	
Malaysia	RM 250	After RM 250 (roughly US$50), PIN is required.
Netherlands	€25	For more than €25 at once or €50 in one day PIN verification is mandatory.
New Zealand	No limit	For each transaction over NZ$80 a PIN is required.
Norway	200 NOK	For each transaction over 200 NOK a PIN is required.
Poland	No limit	For transactions over or equal to 50 PLN PIN is required.
Slovakia	No limit	For transactions over €20 PIN is needed. For every 3 consecutive contactless transactions PIN is needed.
Spain	No limit	For more than €20 PIN verification is mandatory
Sweden	200 SEK	
Switzerland	40 CHF	
Taiwan	No limit	Signatures may be required for large purchases.
Thailand	฿500	
United Kingdom	£30	£20 until 1 September 2015; still £20 limit on some cards. Sometimes PIN is needed anyway to ensure the card is used by its owner. Some retailers will allow higher value purchases using newer hardware that supports high value purchases if the contactless authentication method is biometric (e.g. Apple Touch ID used in Apple Pay)
United States of America	US$25	The actual floor limit is set for each merchant based on the card issuer's risk assessment and other factors. US$25 is the lowest value found, but US$50 is also very common. For lower risk merchants, or those willing to pay a higher "discount rate", a floor limit of US$100 (as found at Costco), or even more, may be assigned.

Machine-readable Passport

A machine-readable passport (MRP) is a machine-readable travel document (MRTD) with the data on the identity page encoded in optical character recognition format. Many countries began to issue machine-readable travel documents in the 1980s.

Page of a passport with machine-readable zone in the red oval

Most travel passports worldwide are MRPs. They are standardized by the ICAO Document 9303 (endorsed by the International Organization for Standardization and the International Electrotechnical Commission as ISO/IEC 7501-1) and have a special machine-readable zone (MRZ), which is usually at the bottom of the identity page at the beginning of a passport. The ICAO Document 9303 describes three types of documents. Usually passport booklets are issued in "Type 3" format, while identity cards and passport cards typically use the "Type 1" format. The machine-readable zone of a Type 3 travel document spans two lines, and each line is 44 characters long. The following information has to be provided in the zone: name, passport number, nationality, date of birth, sex, passport expiration date and personal identity number. There is room for optional, often country-dependent, supplementary information. The machine-readable zone of a Type 1 travel document spans three lines, and each line is 30 characters long.

The advantages of machine-readable passports include:

- Faster processing of arriving passengers by immigration officials.

- More reliable than a human read, compared to the manually read passports that preceded them.

Format

Passport Booklets

Passport booklets have an identity page containing the identity data. This page shall be in the TD3 size which means 125 × 88 mm (4.92 × 3.46 in).

Colombian sample of machine-readable passport

The data of the machine-readable zone consists of two rows of 44 characters each. The only characters used are A–Z, 0–9 and the filler character <.

The format of the first row is:

Positions	Length	Characters	Meaning
1	1	alpha	P, indicating a passport

2	1	alpha+<	Type (for countries that distinguish between different types of passports)
3–5	3	alpha+<	Issuing country or organization (ISO 3166-1 alpha-3 code with modifications)
6–44	39	alpha+<	Surname, followed by two filler characters, followed by given names. Given names are separated by single filler characters

In the name field, spaces, hyphens and other punctuation are represented by <, except apostrophes, which are skipped. If the names are too long, names are abbreviated to their most significant parts. In that case, the last position must contain an alphabetic character to indicate possible truncation, and if there is a given name, the two fillers and at least one character of it must be included.

The format of the second row is:

Posi-tions	Length	Characters	Meaning
1–9	9	alpha+num+<	Passport number
10	1	numeric	Check digit over digits 1–9
11–13	3	alpha+<	Nationality (ISO 3166-1 alpha-3 code with modifications)
14–19	6	numeric	Date of birth (YYMMDD)
20	1	num	Check digit over digits 14–19
21	1	alpha+<	Sex (M, F or < for male, female or unspecified)
22–27	6	numeric	Expiration date of passport (YYMMDD)
28	1	numeric	Check digit over digits 22–27
29–42	14	alpha+num+<	Personal number (may be used by the issuing country as it desires)
43	1	numeric	Check digit over digits 29–42 (may be < if all characters are <)
44	1	numeric	Check digit over digits 1–10, 14–20, and 22–43

The check digit calculation is as follows: each position is assigned a value; for the digits 0 to 9 this is the value of the digits, for the letters A to Z this is 10 to 35, for the filler < this is 0. The value of each position is then multiplied by its weight; the weight of the first position is 7, of the second it is 3, and of the third it is 1, and after that the weights repeat 7, 3, 1, and so on. All values are added together and the remainder of the final value divided by 10 is the check digit.

Some values that are different from ISO 3166-1 alpha-3 are used for the issuing country and nationality field:

- D: Germany

- GBD: British Overseas Territories Citizen (BOTC) (note: the country code of the overseas territory is used to indicate issuing authority and nationality of BOTC), formerly British Dependent Territories Citizen (BDTC)

- GBN: British National (Overseas)

- GBO: British Overseas Citizen

- GBP: British Protected Person

- GBS: British Subject

- PHL: Filipinos

- UNA: specialized agency of the United Nations

- UNK: Resident of Kosovo to whom a travel document has been issued by the United Nations Interim Administration Mission in Kosovo (UNMIK)

- UNO: United Nations organization

- XOM: Sovereign Military Order of Malta

- XXA: Stateless person, as per the 1954 Convention Relating to the Status of Stateless Persons

- XXB: Refugee, as per the 1951 Convention Relating to the Status of Refugees

- XXC: Refugee, other than defined above

- XXX: Unspecified nationality

Other values, which do not have broad acceptance internationally, include:

- WSA: World Service Authority World Passport

Official Travel Documents

Smaller documents such as identity and passport cards are usually in the TD1 size, which is 85.6 * 54.0 mm (3.37 * 2.13 in), the same size as credit cards. The data of the machine-readable zone in a TD1 size card consists of three rows of 30 characters each. The only characters used are A–Z, 0–9 and the filler character <.

Some official travel documents are in the larger TD2 size, 105.0 * 74.0 (4.13 * 2.91 in). They have a layout of the MRZ with two rows of 36 characters each, similar to the TD3 format, but with 31 characters for the name, 7 for the personal number and one less check digit. Yet some official travel documents are in the booklet format with a TD3 identity page.

The format of the first row for TD1 (credit card size) documents is:

Positions	Length	Chars	Meaning
1	1	alpha	I, A or C
2	1	alpha+<	Type, This is at the discretion of the issuing state or authority, but 1–2 should be IP for passport cards, AC for Crew Member Certificates and V is not allowed as 2nd character. ID or I< are typically used for nationally issued ID cards
3–5	3	alpha+<	Issuing country or organization (ISO 3166-1 alpha-3 code with modifications)
6–14	9	alpha+num+<	Document number
15	1	num+<	Check digit over digits 6–14
16–30	15	alpha+num+<	Optional

In addition to ISO 3166-1 alpha-3 code with modifications used for issuing country in passports, also the following organization is accepted:

- XCC Caribbean Community

The format of the second row is:

Positions	Length	Chars	Meaning
1–6	6	num	Date of birth (YYMMDD)
7	1	num	Check digit over digits 1–6
8	1	alpha+<	Sex (M, F or < for male, female or unspecified)
9-14	6	num	Expiration date of document (YYMMDD)
15	1	num	Check digit over digits 9–14
16–18	3	alpha+<	Nationality
19–29	11	alpha+num+<	Optional[1]
30	1	num	Check digit over digits 6–30 (upper line), 1–7, 9–15, 19–29 (middle line)

1: United States Passport Cards, as of 2011, use this field for the application number that produced the card.

The format of the third row is:

Positions	Length	Chars	Meaning
1–30	30	alpha+<	Surname, followed by two filler characters, followed by given names

Machine-readable Visas

The ICAO Document 9303 part 7 describes machine-readable visas. They come in two different formats:

- MRV-A - 80 mm × 120 mm (3.15 in × 4.72 in)

- MRV-B - 74 mm × 105 mm (2.91 in × 4.13 in)

The format of the first row of the machine-readable zone is:

Positions	Length	Chars	Meaning
1	1	alpha	"V"
2	1	alpha+<	Type, this is at the discretion of the issuing state or authority
3–5	3	alpha+<	Issuing country or organization (ISO 3166-1 alpha-3 code with modifications)
6–44	39	alpha+<	Name in MRV-A
6–36	31	alpha+<	Name in MRV-B

The format of the second row is:

Positions	Length	Chars	Meaning
1-9	9	alpha+num+<	Passport or Visa number
10	1	num	Check digit
11–13	3	alpha+<	Nationality
14–19	6	num	Date of birth
20	1	num	Check digit
21	1	alpha+<	Sex
22-27	6	num	Valid until
28	1	num	Check digit
29–44	16	alpha+num+<	Optional data in MRV-A
29–36	8	alpha+num+<	Optional data in MRV-B

Specifications Common to All Formats

The ICAO document 9303 part 3 describes specifications common to all Machine Readable Travel Documents.

The dimensions of the effective reading zone (ERZ) is standardized at 17.0mm (0.67 in) in height with a margin of 3mm at the document edges and 3.2mm at the edge against the visual readable part. This is in order to allow use of a single machine reader.

Only characters A to Z (upper case), 0–9, and < (angle bracket) are allowed.

Nationality Codes and Checksum Calculation

The nationality codes shall contain the ISO 3166-1 alpha-3 code with modifications for all formats, as described in the passport booklets chapter. The check digit calculation method is also the same for all formats.

Names

People's names contain various characters, but must in the Machine Readable Zone (MRZ) be restricted to A–Z and angle brackets.

Apostrophes and similar must be omitted, but hyphens and spaces should be replaced by an angle bracket. Diacritical marks are not permitted in the MRZ. Even though they may be useful to distinguish names, the use of diacritical marks in the MRZ could confuse machine-reading equipment.

Section 6 of the 9303 part 3 document specifies transliteration of letters outside the A–Z range. It recommends that diacritical marks on Latin letters A-Z are simply omitted (ç → C, ð → D, ê → E, etc.), but it allows the following transliterations:

$$å → AA$$

$$ä → AE$$

$$ð → DH$$

ij (Dutch letter; capital form: IJ, the J as part of the ligature being capitalized, too)→ IJ

$$ö → OE$$

$$ü → UE \text{ (German) or UXX (Spanish)}$$
$$ñ → NXX$$

The following transliterations are mandatory:

$$æ → AE$$

$$ø, œ → OE$$

$$ß → SS$$

$$þ → TH$$

There are also tables for the transliteration of names written using Cyrillic and Arabic scripts.

People having names using the listed letters sometimes have trouble with ignorant officials; for example, the document is thought to be a forgery or with airline tickets not having the same spelling as the passport. Consequently, it is often best to use the exact spelling used in the machine-readable zone for the airline ticket or ESTA, and refer to this zone if being asked questions.

Different Spellings of the Same Name Within the Same Document

Names containing non-English letters are usually spelled in the correct way in the non-machine-readable zone of the passport, but are mapped according to the standards of ICAO in the machine-readable zone.

In Germany, Austria and Scandinavia it is standard to use the Å→AA, Ä or Æ→AE, Ö or Ø→OE, Ü→UE, and ß→SS mappings, so Müller becomes MUELLER, Gößmann becomes GOESSMANN, and Hämäläinen becomes HAEMAELAEINEN.

Names originally written in a non-Latin writing systems may pose another problem if there are various internationally recognized transcription standards. For example, the Russian surname Горбачёв is transcribed "Gorbachev" in English and according to the ICAO 9303 rules, "Gorbatschow" in German,"Gorbatchov" in French, "Gorbachov" in Spanish, "Gorbaczow" in Polish, and so on.

Sometimes, as with US visas, simple letters stripped of their proper diacriticals are used (ag: MULLER, GOSSMANN, HAMALAINEN). German credit cards use either the correct or the mapped spelling in their non-machine-readable zone.

First and Given Names

For airline tickets, visas and more, there is an advice to only use the first name, as written in the passport. This is a problem for people who use their second name (as defined by the order they have in the passport) as their main name in daily speech. It is common, for example in Scandinavia, that the second or even third name is the defined for daily usage, for example the actor Hugh Laurie, whose full name is James Hugh Calum Laurie. Swedish travel agents usually book people using the first and daily name if the first one is not their main name, despite advise to use only the first name. If this is too long, the spelling in the MRZ could be used.

For people using a variant of their first name in daily speech, for example the former US president Bill Clinton whose full name is William Jefferson Clinton, the advice is to spell their name as in the passport.

Chinese and Korean names might pose a challenge too, since the family name is normally written first.

Smartdust

Smartdust is a system of many tiny microelectromechanical systems (MEMS) such as sensors, robots, or other devices, that can detect, for example, light, temperature, vibration, magnetism, or chemicals. They are usually operated on a computer network wirelessly and are distributed over some area to perform tasks, usually sensing through radio-frequency identification. Without an antenna of much greater size the range of tiny smart dust communication devices is measured in a few millimeters and they may be vulnerable to electromagnetic disablement and destruction by microwave exposure.

Design and Engineering

The concepts for Smart Dust emerged from a workshop at RAND in 1992 and a series of DARPA ISAT studies in the mid-1990s due to the potential military applications of the technology. The work was strongly influenced by work at UCLA and the University of Michigan during that period, as well as science fiction authors Stanislaw Lem, Neal Stephenson and Vernor Vinge. The first public presentation of the concept by that name was at the American Vacuum Society meeting in Anaheim in 1996.

A Smart Dust research proposal was presented to DARPA written by Kristofer S. J. Pister, Joe Kahn, and Bernhard Boser, all from the University of California, Berkeley, in 1997. The proposal, to build wireless sensor nodes with a volume of one cubic millimeter, was selected for funding in 1998. The project led to a working mote smaller than a grain of rice, and larger "COTS Dust" devices kicked off the TinyOS effort at Berkeley.

The concept was later expanded upon by Kris Pister in 2001. A recent review discusses various techniques to take smartdust in sensor networks beyond millimeter dimensions to the micrometre level.

The Ultra-Fast Systems component of the Nanoelectronics Research Centre at the University of Glasgow is a founding member of a large international consortium which is developing a related concept: smart specks.

Smart dust entered the 2013 Gartner Hype Cycle on emerging technologies as the most speculative entrant.

Transponder Timing

Transponder timing (also called chip timing or RFID timing) is a technique for measuring performance in sport events. A transponder working on a radio-frequency identification (RFID) basis is attached to the athlete and emits a unique code that is detected by radio receivers located at the strategic points in an event.

A finish line which uses transponder timing and RFID technology through overhead antennas and passive, disposable chips.

A ChronoTrack race controller with RFID antennas for detecting transponders attached to runner's shoes.

Prior to the use of this technology, races were either timed by hand (with operators pressing a stopwatch) or using video camera systems.

Transponder Systems

Generically, there are two types of transponder timing systems; active and passive. An active transponder consists of a battery-powered transceiver, connected to the athlete, that emits its unique code when it is interrogated.

A passive transponder does not contain a power source inside the transponder. Instead, the transponder captures electromagnetic energy produced by a nearby exciter and utilizes that energy to emit a unique code.

In both systems, an antenna is placed at the start, finish, and in some cases, intermediate time points and is connected to a decoder. This decoder identifies the unique transponder code and calculates the exact time when the transponder passes a timing point. Some implementations of timing systems require the use of a mat on the ground at the timing points while other systems implement the timing points with vertically oriented portals.

History

RFID was first used in the late 1980s primarily for motor racing and became more widely adopted in athletic events in the mid-1990s on release of low cost 134 kHz transponders and readers from Texas Instruments. This technology formed the basis of electronic sports timing for the world's largest running events as well as for cycling, triathlon and skiing. Some manufacturers made improvements to the technology to handle larger numbers of transponders in the read field or improve the tolerance of their systems to low frequency noise. These low frequency systems are still used a lot today. Other manufacturers developed their own proprietary RFID systems usually as an offshoot to more industrial applications. These latter systems attempted to get around the problem of reading large numbers of transponders in a read field by using the High Frequency 13.56 MHz RFID methodology that allowed transponders to use anti-collision algorithms to avoid tags interfering with each other's signal during the down-link between transponder and reader. Active transponder systems continued to mature and despite their much higher cost they retained market share in the high speed sports like motor racing, cycling and ice skating. Active systems are also used at high profile events such as the Olympics due to their very high read rates and time-stamping precision.

By 2005 a newer RFID technology was becoming available, mostly for industrial applications. The first and second generation (UHF) transponders and readers that were being developed followed a strict protocol to ensure that multiple transponders and readers could be used between manufacturers. Much like the HF tags, the UHF tags were much cheaper to produce in volume and formed the basis in the next revolution in sports timing. Currently many of the largest athletic events are timed using disposable transponders either placed on the back of a race number or on the runners shoe. The low cost meant that transponders were now fully disposable and did not need to be returned to the organizers after the event.

Usage

Very large running events (more than 10,000) and triathlons were the first events to be transponder (or chip) timed because it is near impossible to manually time them. Also for large runs there are delays in participants reaching the start line, which penalize their performance. Some races place antennas or timing mats at both the start line and the finish line, which allow the exact net time to be calculated. Awards in a race are generally based on the "gun time" (which ignores any delay at the start) as per IAAF and USA Track and Field rules. However, some races use "net time" for presenting age group awards.

In the past the transponder was almost always worn on the athletes running shoe, or on an ankle band. This enabled the transponder to be read best on antenna mats because the distance between the transponder and readers antenna is minimized offering the best capture rate. Transponders may be threaded onto the shoe laces for running. For triathlon a soft elastic ankle band holds the transponder to the leg and care is taken to ensure the transponder is in the correct orientation or polarity for maximum read performance. Transponders have also been placed on the race bib. In the past 5 years the newer UHF systems use transponders placed on the shoe lace, or stuck to the race number bib. In both cases, care must be taken to ensure the UHF tag does not directly touch a large part of the skin as this affects read performance. Despite this, UHF Systems have read performances as good (if not better) than the conventional low and high frequency systems. Because these UHF tags are made in huge volumes for industrial applications, their price is much lower than that of conventional re-usable transponders and the race does not bother to collect them afterwards.

Disposable bib with two passive timing chips at the back

All RFID timing systems incorporate a box housing the reader(s) with peripherals like a microprocessor, serial or ethernet communications and power source (battery). The readers are attached to one or more antennas that are designed for the particular operating frequency. In the case of low or medium frequencies these consist of wire loops incorporated into mats that cover the entire width of the timing point. For UHF systems the antennas consist of patch antennas that are protected in a matting system. The patch antennas may also be placed on stands or a finish gantry pointing towards the oncoming athlete. In most cases the distance between reader and antennas is restricted. Also more equipment is needed for events that require multiple timing points. Wider timing points require more readers and antennas. For active systems a simple wire loop is all that is needed since the transponder has its own power source and the loop serves as a trigger to turn on the transponder, then receive the relatively strong signal from the transponder. Therefore, active systems need less readers (or decoders) per timing point width.

Back side of disposable RFID tag used for race timing.

All systems utilize specialised software to calculate results and splits. This software usually resides on a separate PC computer that is connected to the readers via serial or ethernet communications. The software relates the raw transponder code and timestamp data to each entrant in a database and calculates gun and net times of runners, or the splits of a triathlete. In advanced systems these results are instantly calculated and published to the internet so that athletes and spectators have access to results via any web enabled device.

Telecommunication

Telecommunication is the transmission of signs, signals, writings, images and sounds or intelligence of any nature by wire, radio, optical or other electromagnetic systems, as defined by the International Telecommunication Union (ITU).

Earth station at the satellite communication facility in Raisting, Bavaria, Germany

Telecommunication occurs when the exchange of information between communication participants includes the use of technology. It is transmitted either electrically over physical media, such as cables, or via electromagnetic radiation. Such transmission paths are often divided into communication channels which afford the advantages of multiplexing. The term is often used in its plural form, telecommunications, because it involves many different technologies.

Visualization from the Opte Project of the various routes through a portion of the Internet

Early means of communicating over a distance included visual signals, such as beacons, smoke signals, semaphore telegraphs, signal flags, and optical heliographs. Other examples of pre-modern long-distance communication included audio messages such as coded drumbeats, lung-blown horns, and loud whistles. Modern technologies for long-distance communication usually involve electrical and electromagnetic technologies, such as telegraph, telephone, and teleprinter, networks, radio, microwave transmission, fiber optics, and communications satellites.

A revolution in wireless communication began in the first decade of the 20th century with the pioneering developments in radio communications by Guglielmo Marconi, who won the Nobel Prize in Physics in 1909. Other notable pioneering inventors and developers in the field of electrical and electronic telecommunications include Charles Wheatstone and Samuel Morse (telegraph), Alexander Graham Bell (telephone), Edwin Armstrong, and Lee de Forest (radio), as well as Vladimir K. Zworykin, John Logie Baird and Philo Farnsworth (television).

Etymology

The word telecommunication was adapted from the French. It is a compound of the Greek prefix tele- (τηλε-), meaning «distant», and the Latin communicare, meaning "to share"in 1904 by the French engineer and novelist Édouard Estaunié.

History

Beacons and Pigeons

In the Middle Ages, chains of beacons were commonly used on hilltops as a means of relaying a signal. Beacon chains suffered the drawback that they could only pass a single bit of information, so the meaning of the message such as "the enemy has been sighted" had to be agreed upon in advance. One notable instance of their use was during the Spanish Armada, when a beacon chain relayed a signal from Plymouth to London.

A replica of one of Chappe's semaphore towers

In 1792, Claude Chappe, a French engineer, built the first fixed visual telegraphy system (or semaphore line) between Lille and Paris. However semaphore suffered from the need for skilled operators and expensive towers at intervals of ten to thirty kilometres (six to nineteen miles). As a result of competition from the electrical telegraph, the last commercial line was abandoned in 1880.

Homing pigeons have occasionally been used throughout history by different cultures. Pigeon post is thought to have Persians roots and was used by the Romans to aid their military. Frontinus said that Julius Caesar used pigeons as messengers in his conquest of Gaul. The Greeks also conveyed the names of the victors at the Olympic Games to various cities using homing pigeons. In the early 19th century, the Dutch government used the system in Java and Sumatra. And in 1849, Paul Julius Reuter started a pigeon service to fly stock prices between Aachen and Brussels, a service that operated for a year until the gap in the telegraph link was closed.

Telegraph and Telephone

Sir Charles Wheatstone and Sir William Fothergill Cooke invented the electric telegraph in 1837. Also, the first commercial electrical telegraph is purported to have been constructed by Wheatstone and Cooke and opened on 9 April 1839. Both inventors viewed their device as "an improvement to the [existing] electromagnetic telegraph" not as a new device.

Samuel Morse independently developed a version of the electrical telegraph that he unsuccessfully demonstrated on 2 September 1837. His code was an important advance over Wheatstone's signaling method. The first transatlantic telegraph cable was successfully completed on 27 July 1866, allowing transatlantic telecommunication for the first time.

The conventional telephone was invented independently by Alexander Bell and Elisha Gray in 1876. Antonio Meucci invented the first device that allowed the electrical transmission of voice over a line in 1849. However Meucci's device was of little practical value because it relied upon the electrophonic effect and thus required users to place the receiver in their mouth to "hear" what was being said. The first commercial telephone services were set-up in 1878 and 1879 on both sides of the Atlantic in the cities of New Haven and London.

Radio and Television

In 1832, James Lindsay gave a classroom demonstration of wireless telegraphy to his students. By 1854, he was able to demonstrate a transmission across the Firth of Tay from Dundee, Scotland to Woodhaven, a distance of two miles (3 km), using water as the transmission medium. In December 1901, Guglielmo Marconi established wireless communication between St. John's, Newfoundland (Canada) and Poldhu, Cornwall (England), earning him the 1909 Nobel Prize in physics (which he shared with Karl Braun). However small-scale radio communication had already been demonstrated in 1893 by Nikola Tesla in a presentation to the National Electric Light Association.

On 25 March 1925, John Logie Baird was able to demonstrate the transmission of moving pictures at the London department store Selfridges. Baird's device relied upon the Nipkow disk and thus became known as the mechanical television. It formed the basis of experimental broadcasts done by the British Broadcasting Corporation beginning 30 September

1929. However, for most of the twentieth century televisions depended upon the cathode ray tube invented by Karl Braun. The first version of such a television to show promise was produced by Philo Farnsworth and demonstrated to his family on 7 September 1927.

Computers and The Internet

On 11 September 1940, George Stibitz was able to transmit problems using teletype to his Complex Number Calculator in New York and receive the computed results back at Dartmouth College in New Hampshire. This configuration of a centralized computer or mainframe with remote dumb terminals remained popular throughout the 1950s. However, it was not until the 1960s that researchers started to investigate packet switching — a technology that would allow chunks of data to be sent to different computers without first passing through a centralized mainframe. A four-node network emerged on 5 December 1969; this network would become ARPANET, which by 1981 would consist of 213 nodes.

ARPANET development centered around the Request for Comment process and on 7 April 1969, RFC 1 was published. This process is important because ARPANET eventually merged with other networks to form the Internet and many of the protocols the Internet relies upon today were specified through the Request for Comment process. In September 1981, RFC 791 introduced the Internet Protocol v4 (IPv4) and RFC 793 introduced the Transmission Control Protocol (TCP) — thus creating the TCP/IP protocol that much of the Internet relies upon today.

However, not all important developments were made through the Request for Comment process. Two popular link protocols for local area networks (LANs) also appeared in the 1970s. A patent for the token ring protocol was filed by Olof Soderblom on 29 October 1974 and a paper on the Ethernet protocol was published by Robert Metcalfe and David Boggs in the July 1976 issue of Communications of the ACM.

Key Concepts

A number of key concepts reoccur throughout the literature on modern telecommunication theory and systems. Some of these concepts are discussed below.

Basic Elements

Telecommunications is primarily divided up between wired and wireless subtypes. Overall though, a basic telecommunication system consists of three main parts that are always present in some form or another:

- A transmitter that takes information and converts it to a signal.

- A transmission medium, also called the "physical channel" that carries the signal. An example of this is the "free space channel".

- A receiver that takes the signal from the channel and converts it back into usable information for the recipient.

For example, in a radio broadcasting station the station's large power amplifier is the transmitter; and the broadcasting antenna is the interface between the power amplifier and the "free space channel". The free space channel is the transmission medium; and the receiver's antenna is the interface between the free space channel and the receiver. Next, the radio receiver is the destination of the radio signal, and this is where it is converted from electricity to sound for people to listen to.

Sometimes, telecommunication systems are "duplex" (two-way systems) with a single box of electronics working as both the transmitter and a receiver, or a transceiver. For example, a cellular telephone is a transceiver. The transmission electronics and the receiver electronics within a transceiver are actually quite independent of each other. This can be readily explained by the fact that radio transmitters contain power amplifiers that operate with electrical powers measured in watts or kilowatts, but radio receivers deal with radio powers that are measured in the microwatts or nanowatts. Hence, transceivers have to be carefully designed and built to isolate their high-power circuitry and their low-power circuitry from each other, as to not cause interference.

Telecommunication over fixed lines is called point-to-point communication because it is between one transmitter and one receiver. Telecommunication through radio broadcasts is called broadcast communication because it is between one powerful transmitter and numerous low-power but sensitive radio receivers.

Telecommunications in which multiple transmitters and multiple receivers have been designed to cooperate and to share the same physical channel are called multiplex systems. The sharing of physical channels using multiplexing often gives very large reductions in costs. Multiplexed systems are laid out in telecommunication networks, and the multiplexed signals are switched at nodes through to the correct destination terminal receiver.

Analog Versus Digital Communications

Communications signals can be sent either by analog signals or digital signals. There are analog communication systems and digital communication systems. For an analog signal, the signal is varied continuously with respect to the information. In a digital signal, the information is encoded as a set of discrete values (for example, a set of ones and zeros). During the propagation and reception, the information contained in analog signals will inevitably be degraded by undesirable physical noise. (The output of a transmitter is noise-free for all practical purposes.) Commonly, the noise in a communication system can be expressed as adding or subtracting from the desirable signal in a completely random way. This form of noise is called additive noise, with the understanding that the noise can be negative or positive at different instants of time. Noise that is not additive noise is a much more difficult situation to describe or analyze, and these other kinds of noise will be omitted here.

On the other hand, unless the additive noise disturbance exceeds a certain threshold, the information contained in digital signals will remain intact. Their resistance to noise represents a key advantage of digital signals over analog signals.

Telecommunication Networks

A telecommunications network is a collection of transmitters, receivers, and communications channels that send messages to one another. Some digital communications networks contain one or more routers that work together to transmit information to the correct user. An analog communications network consists of one or more switches that establish a connection between two or more users. For both types of network, repeaters may be necessary to amplify or recreate the signal when it is being transmitted over long distances. This is to combat attenuation that can render the signal indistinguishable from the noise. Another advantage of digital systems over analog is that their output is easier to store in memory, i.e. two voltage states (high and low) are easier to store than a continuous range of states.

Communication Channels

The term "channel" has two different meanings. In one meaning, a channel is the physical medium that carries a signal between the transmitter and the receiver. Examples of this include the atmosphere for sound communications, glass optical fibers for some kinds of optical communications, coaxial cables for communications by way of the voltages and electric currents in them, and free space for communications using visible light, infrared waves, ultraviolet light, and radio waves. This last channel is called the "free space channel". The sending of radio waves from one place to another has nothing to do with the presence or absence of an atmosphere between the two. Radio waves travel through a perfect vacuum just as easily as they travel through air, fog, clouds, or any other kind of gas.

The other meaning of the term "channel" in telecommunications is seen in the phrase communications channel, which is a subdivision of a transmission medium so that it can be used to send multiple streams of information simultaneously. For example, one radio station can broadcast radio waves into free space at frequencies in the neighborhood of 94.5 MHz (megahertz) while another radio station can simultaneously broadcast radio waves at frequencies in the neighborhood of 96.1 MHz. Each radio station would transmit radio waves over a frequency bandwidth of about 180 kHz (kilohertz), centered at frequencies such as the above, which are called the "carrier frequencies". Each station in this example is separated from its adjacent stations by 200 kHz, and the difference between 200 kHz and 180 kHz (20 kHz) is an engineering allowance for the imperfections in the communication system.

In the example above, the "free space channel" has been divided into communications channels according to frequencies, and each channel is assigned a separate frequency bandwidth in which to broadcast radio waves. This system of dividing the medium into channels according to frequency is called "frequency-division multiplexing". Another term for the same concept is "wavelength-division multiplexing", which is more commonly used in optical communications when multiple transmitters share the same physical medium.

Another way of dividing a communications medium into channels is to allocate each sender a recurring segment of time (a "time slot", for example, 20 milliseconds out of each second), and to allow each sender to send messages only within its own time slot. This method of dividing the medium into communication channels is called "time-division multiplexing" (TDM), and is used in optical fiber communication. Some radio communication systems use TDM within an allocated FDM channel. Hence, these systems use a hybrid of TDM and FDM.

Modulation

The shaping of a signal to convey information is known as modulation. Modulation can be used to represent a digital message as an analog waveform. This is commonly called "keying" – a term derived from the older use of Morse Code in telecommunications – and several keying techniques exist (these include phase-shift keying, frequency-shift keying, and amplitude-shift keying). The "Bluetooth" system, for example, uses phase-shift keying to exchange information between various devices. In addition, there are combinations of phase-shift keying and amplitude-shift keying which is called (in the jargon of the field) "quadrature amplitude modulation" (QAM) that are used in high-capacity digital radio communication systems.

Modulation can also be used to transmit the information of low-frequency analog signals at higher frequencies. This is helpful because low-frequency analog signals cannot be effectively transmitted over free space. Hence the information from a low-frequency analog signal must be impressed into a higher-frequency signal (known as the "carrier wave") before transmission. There are several different modulation schemes available to achieve this [two of the most basic being amplitude modulation (AM) and frequency modulation (FM)]. An example of this process is a disc jockey's voice being impressed into a 96 MHz carrier wave using frequency modulation (the voice would then be received on a radio as the channel "96 FM"). In addition, modulation has the advantage that it may use frequency division multiplexing (FDM).

Society

Telecommunication has a significant social, cultural and economic impact on modern society. In 2008, estimates placed the telecommunication industry's revenue at $4.7 trillion or just under 3 percent of the gross world product (official exchange rate). Several following sections discuss the impact of telecommunication on society.

Economic Impact

Microeconomics

On the microeconomic scale, companies have used telecommunications to help build global business empires. This is self-evident in the case of online retailer Amazon.com but, according to academic Edward Lenert, even the conventional retailer Walmart has benefited from better telecommunication infrastructure compared to its competitors. In cities throughout the world, home owners use their telephones to order and arrange a variety of home services ranging from pizza deliveries to electricians. Even relatively poor communities have been noted to use telecommunication to their advantage. In Bangladesh's Narshingdi district, isolated villagers use cellular phones to speak directly to wholesalers and arrange a better price for their goods. In Côte d'Ivoire, coffee growers share mobile phones to follow hourly variations in coffee prices and sell at the best price.

Macroeconomics

On the macroeconomic scale, Lars-Hendrik Röller and Leonard Waverman suggested a causal link between good telecommunication infrastructure and economic growth. Few dispute the existence of a correlation although some argue it is wrong to view the relationship as causal.

Because of the economic benefits of good telecommunication infrastructure, there is increasing worry about the inequitable access to telecommunication services amongst various countries of the world—this is known as the digital divide. A 2003 survey by the International Telecommunication Union (ITU) revealed that roughly a third of countries have fewer than one mobile subscription for every 20 people and one-third of countries have fewer than one land-line telephone subscription for every 20 people. In terms of Internet access, roughly half of all countries have fewer than one out of 20 people with Internet access. From this information, as well as educational data, the ITU was able to compile an index that measures the overall ability of citizens to access and use information and communication technologies. Using this measure, Sweden, Denmark and Iceland received the highest ranking while the African countries Nigeria, Burkina Faso and Mali received the lowest.

Social Impact

Telecommunication has played a significant role in social relationships. Nevertheless, devices like the telephone system were originally advertised with an emphasis on the practical dimensions of the device (such as the ability to conduct business or order home services) as opposed to the social dimensions. It was not until the late 1920s and 1930s that the social dimensions of the device became a prominent theme in telephone advertisements. New promotions started appealing to consumers' emotions, stressing the importance of social conversations and staying connected to family and friends.

Since then the role that telecommunications has played in social relations has become increasingly important. In recent years, the popularity of social networking sites has increased dramatically. These sites allow users to communicate with each other as well as post photographs, events and profiles for others to see. The profiles can list a person's age, interests, sexual preference and relationship status. In this way, these sites can play important role in everything from organising social engagements to courtship.

Prior to social networking sites, technologies like short message service (SMS) and the telephone also had a significant impact on social interactions. In 2000, market research group Ipsos MORI reported that 81% of 15- to 24-year-old SMS users in the United Kingdom had used the service to coordinate social arrangements and 42% to flirt.

Other Impacts

In cultural terms, telecommunication has increased the public's ability to access music and film. With television, people can watch films they have not seen before in their own home without having to travel to the video store or cinema. With radio and the Internet, people can listen to music they have not heard before without having to travel to the music store.

Telecommunication has also transformed the way people receive their news. A survey led in 2006 by the non-profit Pew Internet and American Life Project found that when just over 3,000 people living in the United States were asked where they got their news "yesterday", more people said television or radio than newspapers. The results are summarised in the following table (the percentages add up to more than 100% because people were able to specify more than one source).

Local TV	National TV	Radio	Local paper	Internet	National paper
59%	47%	44%	38%	23%	12%

Telecommunication has had an equally significant impact on advertising. TNS Media Intelligence reported that in 2007, 58% of advertising expenditure in the United States was spent on mediums that depend upon telecommunication. The results are summarised in the following table.

	Internet	Radio	Cable TV	Syndicated TV	Spot TV	Network TV	Newspaper	Magazine	Outdoor	Total
Percent	7.6%	7.2%	12.1%	2.8%	11.3%	17.1%	18.9%	20.4%	2.7%	100%
Dollars	$11.31 billion	$10.69 billion	$18.02 billion	$4.17 billion	$16.82 billion	$25.42 billion	$28.22 billion	$30.33 billion	$4.02 billion	$149 billion

Government

Many countries have enacted legislation which conforms to the International Telecommunication Regulations established by the International Telecommunication Union (ITU), which is the "leading UN agency for information and communication technology issues." In 1947, at the Atlantic City Conference, the ITU decided to "afford international protection to all frequencies registered in a new international frequency list and used in conformity with the Radio Regulation." According to the ITU's Radio Regulations adopted in Atlantic City, all frequencies referenced in the International Frequency Registration Board, examined by the board and registered on the International Frequency List "shall have the right to international protection from harmful interference."

From a global perspective, there have been political debates and legislation regarding the management of telecommunication and broadcasting. The history of broadcasting discusses some debates in relation to balancing conventional communication such as printing and telecommunication such as radio broadcasting. The onset of World War II brought on the first explosion of international broadcasting propaganda. Countries, their governments, insurgents, terrorists, and militiamen have all used telecommunication and broadcasting techniques to promote propaganda. Patriotic propaganda for political movements and colonization started the mid-1930s. In 1936, the BBC broadcast propaganda to the Arab World to partly counter similar broadcasts from Italy, which also had colonial interests in North Africa.

Modern insurgents, such as those in the latest Iraq war, often use intimidating telephone calls, SMSs and the distribution of sophisticated videos of an attack on coalition troops within hours of the operation. "The Sunni insurgents even have their own television station, Al-Zawraa, which while banned by the Iraqi government, still broadcasts from Erbil, Iraqi Kurdistan, even as coalition pressure has forced it to switch satellite hosts several times."

On 10 November 2014, President Obama recommended the Federal Communications Commission reclassify broadband Internet service as a telecommunications service in order to preserve net neutrality.

Modern Media

Worldwide Equipment Sales

According to data collected by Gartner and Ars Technica sales of main consumer's telecommunication equipment worldwide in millions of units was:

Equipment / year	1975	1980	1985	1990	1994	1996	1998	2000	2002	2004	2006	2008
Computers	0	1	8	20	40	75	100	135	130	175	230	280

Cell phones	N/A	N/A	N/A	N/A	N/A	N/A	180	400	420	660	830	1000

Telephone

In a telephone network, the caller is connected to the person they want to talk to by switches at various telephone exchanges. The switches form an electrical connection between the two users and the setting of these switches is determined electronically when the caller dials the number. Once the connection is made, the caller's voice is transformed to an electrical signal using a small microphone in the caller's handset. This electrical signal is then sent through the network to the user at the other end where it is transformed back into sound by a small speaker in that person's handset.

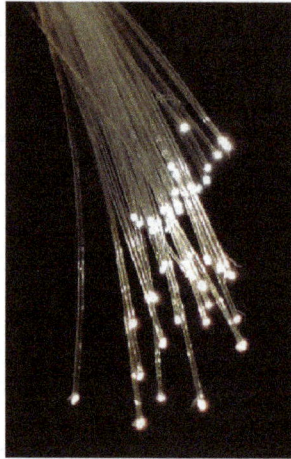

Optical fiber provides cheaper bandwidth for long distance communication.

The landline telephones in most residential homes are analog—that is, the speaker's voice directly determines the signal's voltage. Although short-distance calls may be handled from end-to-end as analog signals, increasingly telephone service providers are transparently converting the signals to digital signals for transmission. The advantage of this is that digitized voice data can travel side-by-side with data from the Internet and can be perfectly reproduced in long distance communication (as opposed to analog signals that are inevitably impacted by noise).

Mobile phones have had a significant impact on telephone networks. Mobile phone subscriptions now outnumber fixed-line subscriptions in many markets. Sales of mobile phones in 2005 totalled 816.6 million with that figure being almost equally shared amongst the markets of Asia/Pacific (204 m), Western Europe (164 m), CEMEA (Central Europe, the Middle East and Africa) (153.5 m), North America (148 m) and Latin America (102 m). In terms of new subscriptions over the five years from 1999, Africa has outpaced other markets with 58.2% growth. Increasingly these phones are being serviced by systems where the voice content is transmitted digitally such as GSM or W-CDMA with many markets choosing to depreciate analog systems such as AMPS.

There have also been dramatic changes in telephone communication behind the scenes. Starting with the operation of TAT-8 in 1988, the 1990s saw the widespread adoption of systems based on optical fibers. The benefit of communicating with optic fibers is that they offer a drastic increase in data capacity. TAT-8 itself was able to carry 10 times as many telephone calls as the last copper cable laid at that time and today's optic fibre cables are able to carry 25 times as many telephone calls as TAT-8. This increase in data capacity is due to several factors: First, optic fibres are physically much smaller than competing technologies. Second, they do not suffer from crosstalk which means several hundred of them can be easily bundled together in a single cable. Lastly, improvements in multiplexing have led to an exponential growth in the data capacity of a single fibre.

Assisting communication across many modern optic fibre networks is a protocol known as Asynchronous Transfer Mode (ATM). The ATM protocol allows for the side-by-side data transmission mentioned in the second paragraph. It is suitable for public telephone networks because it establishes a pathway for data through the network and associates a traffic contract with that pathway. The traffic contract is essentially an agreement between the client and the network about how the network is to handle the data; if the network cannot meet the conditions of the traffic contract it does not accept the connection. This is important because telephone calls can negotiate a contract so as to guarantee themselves a constant bit rate, something that will ensure a caller's voice is not delayed in parts or cut off completely. There are competitors to ATM, such as Multiprotocol Label Switching (MPLS), that perform a similar task and are expected to supplant ATM in the future.

Radio and Television

In a broadcast system, the central high-powered broadcast tower transmits a high-frequency electromagnetic wave to numerous low-powered receivers. The high-frequency wave sent by the tower is modulated with a signal containing visual or audio information. The receiver is then tuned so as to pick up the high-frequency wave and a demodulator is used to retrieve the signal containing the visual or audio information. The broadcast signal can be either analog (signal is varied continuously with respect to the information) or digital (information is encoded as a set of discrete values).

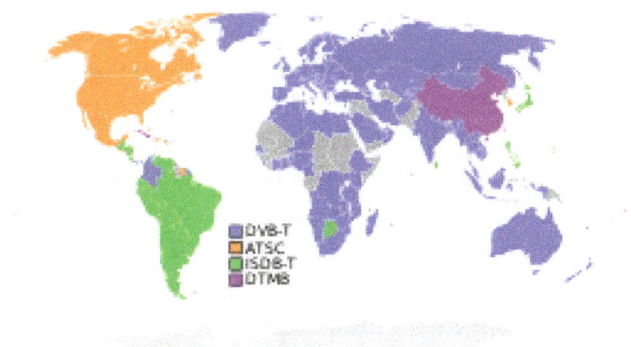

Digital television standards and their adoption worldwide

The broadcast media industry is at a critical turning point in its development, with many countries moving from analog to digital broadcasts. This move is made possible by the production of cheaper, faster and more capable integrated circuits. The chief advantage of digital broadcasts is that they prevent a number of complaints common to traditional analog broadcasts. For television, this includes the elimination of problems such as snowy pictures, ghosting and other distortion. These occur because of the nature of analog transmission, which means that perturbations due to noise will be evident in the final output. Digital transmission overcomes this problem because digital signals are reduced to discrete values upon reception and hence small perturbations do not affect the final output. In a simplified example, if a binary message 1011 was transmitted with signal amplitudes [1.0 0.0 1.0 1.0] and received with signal amplitudes [0.9 0.2 1.1 0.9] it would still decode to the binary message 1011 — a perfect reproduction of what was sent. From this example, a problem with digital transmissions can also be seen in that if the noise is great enough it can significantly alter the decoded message. Using forward error correction a receiver can correct a handful of bit errors in the resulting message but too much noise will lead to incomprehensible output and hence a breakdown of the transmission.

In digital television broadcasting, there are three competing standards that are likely to be adopted worldwide. These are the ATSC, DVB and ISDB standards; the adoption of these standards thus far is presented in the captioned map. All three standards use MPEG-2 for video compression. ATSC uses Dolby Digital AC-3 for audio compression, ISDB uses Advanced Audio Coding (MPEG-2 Part 7) and DVB has no standard for audio compression but typically uses MPEG-1 Part 3 Layer 2. The choice of modulation also varies between the schemes. In digital audio broadcasting, standards are much more unified with practically all countries choosing to adopt the Digital Audio Broadcasting standard (also known as the Eureka 147 standard). The exception is the United States which has chosen to adopt HD Radio. HD Radio, unlike Eureka 147, is based upon a transmission method known as in-band on-channel transmission that allows digital information to "piggyback" on normal AM or FM analog transmissions.

However, despite the pending switch to digital, analog television remains being trans-mitted in most countries. An exception is the United States that ended analog television transmission (by all but the very low-power TV stations) on 12 June 2009 after twice delaying the switchover deadline, Kenya also ended analog television trans-mission in December 2014 after multiple delays. For analog television, there are three standards in use for broadcasting color TV. These are known as PAL (German designed), NTSC (North American designed), and SECAM (French designed). (It is important to understand that these are the ways of sending color TV, and they do not have anything to do with the standards for black & white TV, which also vary from country to country.) For analog radio, the switch to digital radio is made more difficult by the fact that analog receivers are sold at a small fraction of the price of digital receivers. The choice of modulation for analog radio is typically between amplitude (AM) or frequency modulation (FM). To achieve stereo playback, an amplitude modulated subcarrier is used for stereo FM.

Internet

The Internet is a worldwide network of computers and computer networks that communicate with each other using the Internet Protocol. Any computer on the Internet has a unique IP address that can be used by other computers to route information to it. Hence, any computer on the Internet can send a message to any other computer using its IP address. These messages carry with them the originating computer's IP address allowing for two-way communication. The Internet is thus an exchange of messages between computers.

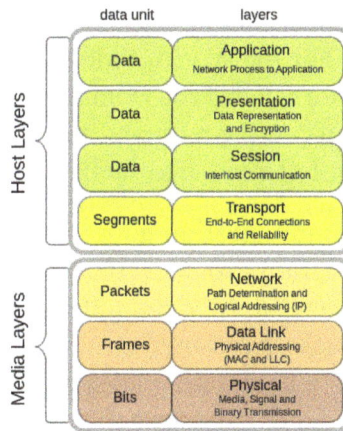

The OSI reference model

It is estimated that the 51% of the information flowing through two-way telecommunications networks in the year 2000 were flowing through the Internet (most of the rest (42%) through the landline telephone). By the year 2007 the Internet clearly dominated and captured 97% of all the information in telecommunication networks (most of the rest (2%) through mobile phones). As of 2008, an estimated 21.9% of the world population has access to the Internet with the highest access rates (measured as a percentage of the population) in North America (73.6%), Oceania/Australia (59.5%) and Europe (48.1%). In terms of broadband access, Iceland (26.7%), South Korea (25.4%) and the Netherlands (25.3%) led the world.

The Internet works in part because of protocols that govern how the computers and routers communicate with each other. The nature of computer network communication lends itself to a layered approach where individual protocols in the protocol stack run more-or-less independently of other protocols. This allows lower-level protocols to be customized for the network situation while not changing the way higher-level protocols operate. A practical example of why this is important is because it allows an Internet browser to run the same code regardless of whether the computer it is running on is connected to the Internet through an Ethernet or Wi-Fi connection. Protocols are often talked about in terms of their place in the OSI reference model (pictured on the right), which emerged in 1983 as the first step in an unsuccessful attempt to build a universally adopted networking protocol suite.

For the Internet, the physical medium and data link protocol can vary several times as packets traverse the globe. This is because the Internet places no constraints on what physical medium or data link protocol is used. This leads to the adoption of media and protocols that best suit the local network situation. In practice, most intercontinental communication will use the Asynchronous Transfer Mode (ATM) protocol (or a modern equivalent) on top of optic fiber. This is because for most intercontinental communication the Internet shares the same infrastructure as the public switched telephone network.

At the network layer, things become standardized with the Internet Protocol (IP) being adopted for logical addressing. For the World Wide Web, these "IP addresses" are derived from the human readable form using the Domain Name System (e.g. 72.14.207.99 is derived from www.google.com). At the moment, the most widely used version of the Internet Protocol is version four but a move to version six is imminent.

At the transport layer, most communication adopts either the Transmission Control Protocol (TCP) or the User Datagram Protocol (UDP). TCP is used when it is essential every message sent is received by the other computer whereas UDP is used when it is merely desirable. With TCP, packets are retransmitted if they are lost and placed in order before they are presented to higher layers. With UDP, packets are not ordered or retransmitted if lost. Both TCP and UDP packets carry port numbers with them to specify what application or process the packet should be handled by. Because certain application-level protocols use certain ports, network administrators can manipulate traffic to suit particular requirements. Examples are to restrict Internet access by blocking the traffic destined for a particular port or to affect the performance of certain applications by assigning priority.

Above the transport layer, there are certain protocols that are sometimes used and loosely fit in the session and presentation layers, most notably the Secure Sockets Layer (SSL) and Transport Layer Security (TLS) protocols. These protocols ensure that data transferred between two parties remains completely confidential. Finally, at the application layer, are many of the protocols Internet users would be familiar with such as HTTP (web browsing), POP3 (e-mail), FTP (file transfer), IRC (Internet chat), Bit-Torrent (file sharing) and XMPP (instant messaging).

Voice over Internet Protocol (VoIP) allows data packets to be used for synchronous voice communications. The data packets are marked as voice type packets and can be prioritized by the network administrators so that the real-time, synchronous conversation is less subject to contention with other types of data traffic which can be delayed (i.e. file transfer or email) or buffered in advance (i.e. audio and video) without detriment. That prioritization is fine when the network has sufficient capacity for all the VoIP calls taking place at the same time and the network is enabled for prioritization i.e. a private corporate style network, but the Internet is not generally managed in this way and so there can be a big difference in the quality of VoIP calls over a private network and over the public Internet.

Local Area Networks and Wide Area Networks

Despite the growth of the Internet, the characteristics of local area networks (LANs)--computer networks that do not extend beyond a few kilometers—remain distinct. This is because networks on this scale do not require all the features associated with larger networks and are often more cost-effective and efficient without them. When they are not connected with the Internet, they also have the advantages of privacy and security. However, purposefully lacking a direct connection to the Internet does not provide assured protection from hackers, military forces, or economic powers. These threats exist if there are any methods for connecting remotely to the LAN.

Wide area networks (WANs) are private computer networks that may extend for thousands of kilometers. Once again, some of their advantages include privacy and security. Prime users of private LANs and WANs include armed forces and intelligence agencies that must keep their information secure and secret.

In the mid-1980s, several sets of communication protocols emerged to fill the gaps between the data-link layer and the application layer of the OSI reference model. These included Appletalk, IPX, and NetBIOS with the dominant protocol set during the early 1990s being IPX due to its popularity with MS-DOS users. TCP/IP existed at this point, but it was typically only used by large government and research facilities.

As the Internet grew in popularity and its traffic was required to be routed into private networks, the TCP/IP protocols replaced existing local area network technologies. Additional technologies, such as DHCP, allowed TCP/IP-based computers to self-configure in the network. Such functions also existed in the AppleTalk/ IPX/ NetBIOS protocol sets.

Whereas Asynchronous Transfer Mode (ATM) or Multiprotocol Label Switching (MPLS) are typical data-link protocols for larger networks such as WANs; Ethernet and Token Ring are typical data-link protocols for LANs. These protocols differ from the former protocols in that they are simpler, e.g., they omit features such as quality of service guarantees, and offer collision prevention. Both of these differences allow for more economical systems.

Despite the modest popularity of IBM Token Ring in the 1980s and 1990s, virtually all LANs now use either wired or wireless Ethernet facilities. At the physical layer, most wired Ethernet implementations use copper twisted-pair cables (including the common 10BASE-T networks). However, some early implementations used heavier coaxial cables and some recent implementations (especially high-speed ones) use optical fibers. When optic fibers are used, the distinction must be made between multimode fibers and single-mode fibers. Multimode fibers can be thought of as thicker optical fibers that are cheaper to manufacture devices for, but that suffers from less usable bandwidth and worse attenuation – implying poorer long-distance performance.

Transmission Capacity

The effective capacity to exchange information worldwide through two-way telecommunication networks grew from 281 petabytes of (optimally compressed) information in 1986, to 471 petabytes in 1993, to 2.2 (optimally compressed) exabytes in 2000, and to 65 (optimally compressed) exabytes in 2007. This is the informational equivalent of two newspaper pages per person per day in 1986, and six entire newspapers per person per day by 2007. Given this growth, telecommunications play an increasingly important role in the world economy and the global telecommunications industry was about a $4.7 trillion sector in 2012. The service revenue of the global telecommunications industry was estimated to be $1.5 trillion in 2010, corresponding to 2.4% of the world's gross domestic product (GDP).

Intelligent Transportation System

Intelligent transportation systems (ITS) are advanced applications which, without embodying intelligence as such, aim to provide innovative services relating to different modes of transport and traffic management and enable various users to be better informed and make safer, more coordinated, and 'smarter' use of transport networks.

ITS graphical user interface displaying the Hungarian highway network and its data points

Although ITS may refer to all modes of transport, EU Directive 2010/40/EU (7 July 2010) defines ITS as systems in which information and communication technologies are applied in the field of road transport, including infrastructure, vehicles and users, and in traffic management and mobility management, as well as for interfaces with other modes of transport.

Background

Recent governmental activity in the area of ITS – specifically in the United States – is further motivated by an increasing focus on homeland security. Many of the proposed

ITS systems also involve surveillance of the roadways, which is a priority of homeland security. Funding of many systems comes either directly through homeland security organisations or with their approval. Further, ITS can play a role in the rapid mass evacuation of people in urban centers after large casualty events such as a result of a natural disaster or threat. Much of the infrastructure and planning involved with ITS parallels the need for homeland security systems.

In the developing world, the migration from rural to urbanized habitats has progressed differently. Many areas of the developing world have urbanised without significant motorisation and the formation of suburbs. A small portion of the population can afford automobiles, but the automobiles greatly increase congestion in these multimodal transportation systems. They also produce considerable air pollution, pose a significant safety risk, and exacerbate feelings of inequities in the society. High-population density could be supported by a multimodal system of walking, bicycle transportation, motorcycles, buses, and trains.

Other parts of the developing world, such as China, India & Brazil remain largely rural but are rapidly urbanising and industrialising. In these areas a motorised infrastructure is being developed alongside motorisation of the population. Great disparity of wealth means that only a fraction of the population can motorise, and therefore the highly dense multimodal transportation system for the poor is cross-cut by the highly motorised transportation system for the rich.

Intelligent Transportation Technologies

Intelligent transport systems vary in technologies applied, from basic management systems such as car navigation; traffic signal control systems; container management systems; variable message signs; automatic number plate recognition or speed cameras to monitor applications, such as security CCTV systems; and to more advanced applications that integrate live data and feedback from a number of other sources, such as parking guidance and information systems; weather information; bridge de-icing (US deicing) systems; and the like. Additionally, predictive techniques are being developed to allow advanced modelling and comparison with historical baseline data. Some of these technologies are described in the following sections.

Wireless Communications

Various forms of wireless communications technologies have been proposed for intelligent transportation systems.

Radio modem communication on UHF and VHF frequencies are widely used for short and long range communication within ITS.

Short-range communications of 350 m can be accomplished using IEEE 802.11 protocols, specifically WAVE or the Dedicated Short Range Communications standard being promoted by the Intelligent Transportation Society of America and the United States Department of Transportation. Theoretically, the range of these protocols can be extended using Mobile ad hoc networks or Mesh networking.

Traffic monitoring gantry with wireless communication dish antenna

Longer range communications have been proposed using infrastructure networks such as WiMAX (IEEE 802.16), Global System for Mobile Communications (GSM), or 3G. Long-range communications using these methods are well established, but, unlike the short-range protocols, these methods require extensive and very expensive infrastructure deployment. There is lack of consensus as to what business model should support this infrastructure.

Auto Insurance companies have utilised ad hoc solutions to support eCall and behavioural tracking functionalities in the form of Telematics 2.0.

Computational Technologies

Recent advances in vehicle electronics have led to a move towards fewer, more capable computer processors on a vehicle. A typical vehicle in the early 2000s would have between 20 and 100 individual networked microcontroller/Programmable logic controller modules with non-real-time operating systems. The current trend is toward fewer, more costly microprocessor modules with hardware memory management and Real-Time Operating Systems. The new embedded system platforms allow for more sophisticated software applications to be implemented, including model-based process control, artificial intelligence, and ubiquitous computing. Perhaps the most important of these for Intelligent Transportation Systems is artificial intelligence.

Floating Car Data/Floating Cellular Data

"Floating car" or "probe" data collection is a set of relatively low-cost methods for obtaining travel time and speed data for vehicles travelling along streets, highways, motorways (freeways), and other transport routes. Broadly speaking, four methods have been used to obtain the raw data:

- Triangulation method. In developed countries a high proportion of cars contain one or more mobile phones. The phones periodically transmit their presence information to the mobile phone network, even when no voice connection is established. In the mid-2000s, attempts were made to use mobile phones as anonymous traffic probes. As a car moves, so does the signal of any mobile phones that are inside the vehicle. By measuring and analysing network data using triangulation, pattern matching or cell-sector statistics (in an anonymous format), the data was converted into traffic flow information. With more congestion, there are more cars, more phones, and thus, more probes. In metropolitan areas, the distance between antennas is shorter and in theory accuracy increases. An advantage of this method is that no infrastructure needs to be built along the road; only the mobile phone network is leveraged. But in practice the triangulation method can be complicated, especially in areas where the same mobile phone towers serve two or more parallel routes (such as a motorway (freeway) with a frontage road, a motorway (freeway) and a commuter rail line, two or more parallel streets, or a street that is also a bus line). By the early 2010s, the popularity of the triangulation method was declining.

- Vehicle re-identification. Vehicle re-identification methods require sets of detectors mounted along the road. In this technique, a unique serial number for a device in the vehicle is detected at one location and then detected again (re-identified) further down the road. Travel times and speed are calculated by comparing the time at which a specific device is detected by pairs of sensors. This can be done using the MAC addresses from Bluetooth or other devices, or using the RFID serial numbers from Electronic Toll Collection (ETC) transponders (also called "toll tags").

- GPS based methods. An increasing number of vehicles are equipped with in-vehicle satnav/GPS (satellite navigation) systems that have two-way communication with a traffic data provider. Position readings from these vehicles are used to compute vehicle speeds. Modern methods may not use dedicated hardware but instead Smartphone based solutions using so called Telematics 2.0 approaches.

- Smartphone-based rich monitoring. Smartphones having various sensors can be used to track traffic speed and density. The accelerometer data from smartphones used by car drivers is monitored to find out traffic speed and road quality. Audio data and GPS tagging of smartphones enables identification of traffic density and possible traffic jams. This was implemented in Bangalore, India as a part of a research experimental system Nericell.

Floating car data technology provides advantages over other methods of traffic measurement:

- Less expensive than sensors or cameras

- More coverage (potentially including all locations and streets)

- Faster to set up and less maintenance

- Works in all weather conditions, including heavy rain

Sensing Technologies

Technological advances in telecommunications and information technology, coupled with ultramodern/state-of-the-art microchip, RFID (Radio Frequency Identification), and inexpensive intelligent beacon sensing technologies, have enhanced the technical capabilities that will facilitate motorist safety benefits for intelligent transportation systems globally. Sensing systems for ITS are vehicle- and infrastructure-based networked systems, i.e., Intelligent vehicle technologies. Infrastructure sensors are indestructible (such as in-road reflectors) devices that are installed or embedded in the road or surrounding the road (e.g., on buildings, posts, and signs), as required, and may be manually disseminated during preventive road construction maintenance or by sensor injection machinery for rapid deployment. Vehicle-sensing systems include deployment of infrastructure-to-vehicle and vehicle-to-infrastructure electronic beacons for identification communications and may also employ video automatic number plate recognition or vehicle magnetic signature detection technologies at desired intervals to increase sustained monitoring of vehicles operating in critical zones.

Inductive Loop Detection

Inductive loops can be placed in a roadbed to detect vehicles as they pass through the loop's magnetic field. The simplest detectors simply count the number of vehicles during a unit of time (typically 60 seconds in the United States) that pass over the loop, while more sophisticated sensors estimate the speed, length, and class of vehicles and the distance between them. Loops can be placed in a single lane or across multiple lanes, and they work with very slow or stopped vehicles as well as vehicles moving at high speed.

Saw cut loop detectors for vehicle detection buried in the pavement at this intersection as seen by the rectangular shapes of loop detector sealant at the bottom part of this picture.

Video Vehicle Detection

Traffic-flow measurement and automatic incident detection using video cameras is another form of vehicle detection. Since video detection systems such as those used in automatic number plate recognition do not involve installing any components directly into the road surface or roadbed, this type of system is known as a "non-intrusive" method of traffic detection. Video from cameras is fed into processors that analyse the changing characteristics of the video image as vehicles pass. The cameras are typically mounted on poles or structures above or adjacent to the roadway. Most video detection systems require some initial configuration to "teach" the processor the baseline background image. This usually involves inputting known measurements such as the distance between lane lines or the height of the camera above the roadway. A single video detection processor can detect traffic simultaneously from one to eight cameras, depending on the brand and model. The typical output from a video detection system is lane-by-lane vehicle speeds, counts, and lane occupancy readings. Some systems provide additional outputs including gap, headway, stopped-vehicle detection, and wrong-way vehicle alarms.

Bluetooth Detection

Bluetooth is an accurate and inexpensive way to measure travel time and make origin and destination analysis. Bluetooth is a wireless standard used to communicate between electrondresses from Bluetooth devices in passing vehicles. If these sensors are interconnected they are able to calculate travel time and provide data for origin and destination matrices. Compared to other traffic measurement technologies, Bluetooth measurement has some differences:

- Accurate measurement points with absolute confirmation to provide to the second travel times.

- Is non-intrusive, which can lead to lower-cost installations for both permanent and temporary sites.

- Is limited to how many Bluetooth devices are broadcasting in a vehicle so counting and other applications are limited.

- Systems are generally quick to set up with little to no calibration needed.

Since Bluetooth devices become more prevalent on board vehicles and with more portable electronics broadcasting, the amount of data collected over time becomes more accurate and valuable for travel time and estimation purposes.

Audio Detection

It is also possible to measure traffic density on a road using the Audio signal that consists of the cumulative sound from tire noise, engine noise, engine-idling noise, honks and air turbulence noise. A roadside-installed microphone picks up the audio that comprises the various vehicle noise and Audio signal processing techniques can be used to estimate the traffic state. The accuracy of such a system compares well with the other methods described above.

Information Fusion from Multiple Traffic Sensing Modalities

The data from the different sensing technologies can be combined in intelligent ways to determine the traffic state accurately. A Data fusion based approach that utilizes the road side collected acoustic, image and sensor data has been shown to combine the advantages of the different individual methods.

Intelligent Transportation Applications

Emergency Vehicle Notification Systems

The in-vehicle eCall is generated either manually by the vehicle occupants or automatically via activation of in-vehicle sensors after an accident. When activated, the in-vehicle eCall device will establish an emergency call carrying both voice and data directly to the nearest emergency point (normally the nearest E1-1-2 Public-safety answering point, PSAP). The voice call enables the vehicle occupant to communicate with the trained eCall operator. At the same time, a minimum set of data will be sent to the eCall operator receiving the voice call.

The minimum set of data contains information about the incident, including time, precise location, the direction the vehicle was traveling, and vehicle identification. The pan-European eCall aims to be operative for all new type-approved vehicles as a standard option. Depending on the manufacturer of the eCall system, it could be mobile phone based (Bluetooth connection to an in-vehicle interface), an integrated eCall device, or a functionality of a broader system like navigation, Telematics device, or tolling device. eCall is expected to be offered, at earliest, by the end of 2010, pending standardization by the European Telecommunications Standards Institute and commitment from large EU member states such as France and the United Kingdom.

Congestion pricing gantry at North Bridge Road, Singapore.

The EC funded project SafeTRIP is developing an open ITS system that will improve road safety and provide a resilient communication through the use of S-band satellite communication. Such platform will allow for greater coverage of the Emergency Call Service within the EU.

Automatic Road Enforcement

A traffic enforcement camera system, consisting of a camera and a vehicle-monitoring device, is used to detect and identify vehicles disobeying a speed limit or some other road legal requirement and automatically ticket offenders based on the license plate number. Traffic tickets are sent by mail. Applications include:

Automatic speed enforcement gantry or "Lombada Eletrônica" with ground sensors at Brasilia, D.F.

- Speed cameras that identify vehicles traveling over the legal speed limit. Many such devices use radar to detect a vehicle's speed or electromagnetic loops buried in each lane of the road.

- Red light cameras that detect vehicles that cross a stop line or designated stopping place while a red traffic light is showing.

- Bus lane cameras that identify vehicles traveling in lanes reserved for buses. In some jurisdictions, bus lanes can also be used by taxis or vehicles engaged in car pooling.

- Level crossing cameras that identify vehicles crossing railways at grade illegally.

- Double white line cameras that identify vehicles crossing these lines.

- High-occupancy vehicle lane cameras that identify vehicles violating HOV requirements.

- Turn cameras at intersections where specific turns are prohibited on red. This type of camera is mostly used in cities or heavy populated areas.

Variable Speed Limits

Recently some jurisdictions have begun experimenting with variable speed limits that change with road congestion and other factors. Typically such speed limits only change to decline during poor conditions, rather than being improved in good ones. One example is on Britain's M25 motorway, which circumnavigates London. On the most heavily traveled 14-mile (23 km) section (junction 10 to 16) of the M25 variable speed limits combined with automated enforcement have been in force since 1995. Initial results indicated savings in journey times, smoother-flowing traffic, and a fall in the number of accidents, so the implementation was made permanent in 1997. Further trials on the M25 have been thus far proven inconclusive.

Example variable speed limit sign in the United States.

Collision Avoidance Systems

Japan has installed sensors on its highways to notify motorists that a car is stalled ahead.

Dynamic Traffic Light Sequence

A 2008 paper was written about using RFID for dynamic traffic light sequences. It circumvents or avoids problems that usually arise with systems that use image processing and beam interruption techniques. RFID technology with appropriate algorithm and database were applied to a multi-vehicle, multi-lane and multi-road junction area to provide an efficient time management scheme. A dynamic time schedule was worked out for the passage of each column. The simulation showed the dynamic sequence algorithm could adjust itself even with the presence of some extreme cases. The paper said the system could emulate the judgment of a traffic police officer on duty, by considering the number of vehicles in each column and the routing proprieties.

Cooperative Systems on the Road

Communication cooperation on the road includes car-to-car, car-to-infrastructure, and vice versa. Data available from vehicles are acquired and transmitted to a server for central fusion and processing. These data can be used to detect events such as rain (wiper activity) and congestion (frequent braking activities). The server processes a driving recommendation dedicated to a single or a specific group of drivers and transmits it wirelessly to vehicles. The goal of cooperative systems is to use and plan communication and sensor infrastructure to increase road safety. The definition of cooperative systems in road traffic is according to the European Commission:

> "Road operators, infrastructure, vehicles, their drivers and other road users will cooperate to deliver the most efficient, safe, secure and comfortable journey. The vehicle-vehicle and vehicle-infrastructure co-operative systems will contribute to these objectives beyond the improvements achievable with stand-alone systems."

ITS World Congress is a world-wide annual event to promote and showcase ITS technologies. ERTICO – ITS Europe, ITS America and ITS AsiaPacific work closely together in the preparation of the annual ITS World Congress and Exhibition attracting over 8,000 people. Each year the event takes place in a different region (Europe, Americas or Asia-Pacific).

The first ITS World Congress was held in Paris in 1994 and it has become the most important annual ITS event. Subsequent events have been held in:

- 1995 Yokohama
- 1996 Orlando
- 1997 Berlin
- 1998 Seoul
- 1999 Toronto
- 2000 Turin
- 2001 Sydney
- 2002 Chicago
- 2003 Madrid
- 2004 Nagoya
- 2005 San Francisco
- 2006 London
- 2007 Beijing
- 2008 New York
- 2009 Stockholm

- 2010 Busan
- 2011 Orlando
- 2012 Vienna
- 2013 Tokyo
- 2014 Detroit
- 2015 Bordeaux

Future Congress are scheduled for Melbourne 2016, Montreal 2017, Copenhagen 2018, Singapore 2019.

Europe

The Network of National ITS Associations is a grouping of national ITS interests formed in order to ensure that ITS knowledge and information is transmitted to all actors at the local and national level – such as small and medium-sized companies – and support ITS promotion from the ground up. It was officially launched 7 October 2004 in London. The Network currently consists of 27 member organisations. The Network Secretariat is at ERTICO – ITS Europe.

ERTICO – ITS Europe is a multi-sector, public/private partnership pursuing the development and deployment of Intelligent Transport Systems and Services (ITS). They connect public authorities, industry players, infrastructure operators, users, national ITS associations and other organisations together and work to bring "Intelligence into Mobility". The ERTICO work programme focuses on initiatives to improve transport safety, security and network efficiency whilst taking into account measures to reduce environmental impact. Our vision is of a future transport system working towards zero accidents, zero delays and fully informed people, where services are affordable and seamless, the environment is protected, privacy is respected and security is provided.

United States

In the United States, each state has an Intelligent Transportation Systems (ITS) chapter that holds a yearly conference to promote and showcase ITS technologies and ideas. Representatives from each Department of Transportation (state, cities, towns, and counties) within the state attend this conference.

Transponder

In telecommunication, a transponder is one of two types of devices. In air navigation or radio frequency identification, a flight transponder is a device that emits an identifying signal in response to an interrogating received signal. In a communications satellite, a transponder gathers signals over a range of uplink frequencies and re-transmits them on a different set of downlink frequencies to receivers on Earth, often without changing the content of the received signal or signals.

A Highway 407 toll transponder

The term is a portmanteau for transmitter-responder. It is variously abbreviated as XPDR, XPNDR, TPDR or TP.

Satellite/Broadcast Communications

A communications satellite's channels are called transponders, because each is a separate transceiver or repeater. With digital video data compression and multiplexing, several video and audio channels may travel through a single transponder on a single wideband carrier. Original analog video only has one channel per transponder, with subcarriers for audio and automatic transmission identification service (ATIS). Non-multiplexed radio stations can also travel in single channel per carrier (SCPC) mode, with multiple carriers (analog or digital) per transponder. This allows each station to transmit directly to the satellite, rather than paying for a whole transponder, or using landlines to send it to an earth station for multiplexing with other stations.

Optical Communications

In optical fiber communications, a transponder is the element that sends and receives the optical signal from a fiber. A transponder is typically characterized by its data rate and the maximum distance the signal can travel.

The term "transponder" can apply to different items with important functional differences, mentioned across academic and commercial literature:

- according to one description, a transponder and transceiver are both functionally similar devices that convert a full-duplex electrical signal into a full-duplex optical signal. The difference between the two is that transceivers interface electrically with the host system using a serial interface, whereas transponders use a parallel interface to do so. In this view, transponders provide easier-to-handle lower-rate parallel signals, but are bulkier and consume more power than transceivers.

- according to another description, transceivers are limited to providing an electrical-optical function only (not differentiating between serial or parallel electrical interfaces), whereas transponders convert an optical signal at one wavelength to an optical signal at another wavelength (typically ITU standardized for DWDM communication). As such, transponders can be considered as two transceivers placed back-to-back. This view also seems to be held by e.g. Fujitsu.

As a result, difference in transponder functionality also might influence the functional description of related optical modules like transceivers and muxponders.

Aviation

Another type of transponder occurs in identification friend or foe systems in military aviation and in air traffic control secondary surveillance radar (beacon radar) systems for general aviation and commercial aviation. Primary radar works best with large all-metal aircraft, but not so well on small, composite aircraft. Its range is also limited by terrain and rain or snow and also detects unwanted objects such as automobiles, hills and trees. Furthermore, it cannot always estimate the altitude of an aircraft. Secondary radar overcomes these limitations but it depends on a transponder in the aircraft to respond to interrogations from the ground station to make the plane more visible.

Depending on the type of interrogation, the transponder sends back a transponder code (or "squawk code", Mode A) or altitude information (Mode C) to help air traffic controllers to identify the aircraft and to maintain separation between planes. Another mode called Mode S (Mode Select) is designed to help avoiding over-interrogation of the transponder (having many radars in busy areas) and to allow automatic collision avoidance. Mode S transponders are "backwards compatible" with Modes A & C. Mode S is mandatory in controlled airspace in many countries. Some countries have also required, or are moving towards requiring, that all aircraft be equipped with Mode S, even in uncontrolled airspace. However, in the field of general aviation there have been objections to these moves, because of the cost, size, limited benefit to the users in uncontrolled airspace, and, in the case of balloons and gliders, the power requirements during long flights.

Marine

The International Maritime Organization's International Convention for the Safety of Life at Sea requires the Automatic Identification System (AIS) to be fitted aboard international voyaging ships with gross tonnage (GT) of 300 or more, and all passenger ships regardless of size. Although AIS transmitters/receivers are generally called transponders they generally transmit autonomously, although coast stations can interrogate class B transponders on smaller vessels for additional information. In addition, navigational aids often have transponders called RACON (radar beacons) designed to make them stand out on a ship's radar screen.

Automotive

Many modern automobiles have keys with transponders hidden inside the plastic head of the key. The user of the car may not even be aware that the transponder is there, because there are no buttons to press. When a key is inserted into the ignition lock cylinder and turned, the car's computer sends a radio signal to the transponder. Unless the transponder replies with a valid code, the computer will not allow the engine to be started. Transponder keys have no battery; they are energized by the radio signal itself.

Road

Electronic toll collection systems such as E-ZPass in the eastern United States use RFID transponders to identify vehicles. Highway 407 in Ontario is one of the world's first completely automated toll highways.

Motorsport

Transponders are used in motorsport for lap timing purposes. A cable loop is dug into the race circuit near to the start/finish line. Each car has an active transponder with a unique ID code. When the racing car passes the start/finish line the lap time and the racing position is shown on the score board.

Passive and active RFID systems are used in off-road events such as Enduro and Hare and Hounds racing, the riders have a transponder on their person, normally on their arm. When they complete a lap they swipe or touch the receiver which is connected to a computer and log their lap time. The Casimo Group Ltd make a system which does this.

NASCAR uses transponders and cable loops placed at numerous points around the track to determine the lineup during a caution period. This system replaced a dangerous race back to the start-finish line.

Underwater

Sonar transponders operate under water and are used to measure distance and form the basis of underwater location marking, position tracking and navigation.

Gated Communities

Transponders may also be used by residents to enter their gated communities. However, having more than one transponder causes problems. If a resident's car with simple transponder is parked in the vicinity, any vehicle can come up to the automated gate, triggering the gate interrogation signal, which may get an acceptable response from the resident's car. Such units properly installed might involve beamforming, unique transponders for each vehicle, or simply obliging vehicles to be stored away from the gate.

References

- Blechman, Andrew (2007). Pigeons-The fascinating saga of the world's most revered and reviled bird. St Lucia, Queensland: University of Queensland Press. ISBN 9780702236419.

- Ambardar, Ashok (1999). Analog and Digital Signal Processing (2nd ed.). Brooks/Cole Publishing Company. pp. 1–2. ISBN 0-534-95409-X.

- Stallings, William (2004). Data and Computer Communications (7th edition (intl) ed.). Pearson Prentice Hall. pp. 337–366. ISBN 0-13-183311-1.

- Martin, Michael (2000). Understanding the Network (The Networker's Guide to AppleTalk, IPX, and NetBIOS), SAMS Publishing, ISBN 0-7357-0977-7.

- Hafner, Katie (1998). Where Wizards Stay Up Late: The Origins Of The Internet. Simon & Schuster. ISBN 0-684-83267-4.

- "Todo lo que necesitas saber sobre el Pasaporte Electrónico" (in Spanish). Superintendencia Nacional de Migraciones. Retrieved 22 February 2016.

- "Visa Schengen: su eliminación estará lista para marzo del 2016" (in Spanish). El Comercio. 30 December 2015. Retrieved 23 February 2016.

- Coren, Michael J. (2011-12-05). "VA's Real-Time Location System: A way to improve patient safety, or Big Brother?". Nextgov.com. Retrieved 2016-04-28.

- "How RF Controls Technology Paves the Way for the "Internet of Everything." | RF Controls". Rfctrls.com. 2014-05-07. Retrieved 2016-04-28.

- "RFID Technology from Texas Instruments and RF Code Brings Service and Safety to Guests at Steamboat Ski Resort" (PDF). Rfidjournalevents.com. Retrieved 2016-04-28.

- "IEEE Xplore Abstract - Enhancing Accuracy Performance of Bluetooth Positioning". Ieeexplore. ieee.org. 2007-03-15. doi:10.1109/WCNC.2007.506. Retrieved 2016-04-28.

- "JAMA Network | JAMA | Electromagnetic Interference From Radio Frequency Identification Inducing Potentially Hazardous Incidents in Critical Care Medical Equipment". Jama.jamanetwork.com. Retrieved 2016-04-28.

- "IEEE Xplore Abstract - WLAN location determination via clustering and probability distributions". Ieeexplore.ieee.org. 2003-03-26. doi:10.1109/PERCOM.2003.1192736. Retrieved 2016-04-28.

- Campbell, Francis. "Contactless payments taking off in the UK in 2015". mobiletransaction.org. Retrieved 16 December 2015.

- Wyatt, Edward (10 November 2014). "Obama Asks F.C.C. to Adopt Tough Net Neutrality Rules". New York Times. Retrieved 15 November 2014.

- NYT Editorial Board (14 November 2014). "Why the F.C.C. Should Heed President Obama on Internet Regulation". New York Times. Retrieved 15 November 2014.

6

Radio Navigation: Diverse Dimensions

Radio navigation is used by flights and ships worldwide to receive position information transmitted from ground stations. Long-range radio navigation is not very accurate but short-range navigation can provide precise location within a few meters. The content covers topics like radiolocation, direction finding and fuzzy locating system. Radio navigation is best understood in confluence with the major topics listed in the following chapter.

Radio Navigation

The basic principles are measurements from/to electric beacons, especially

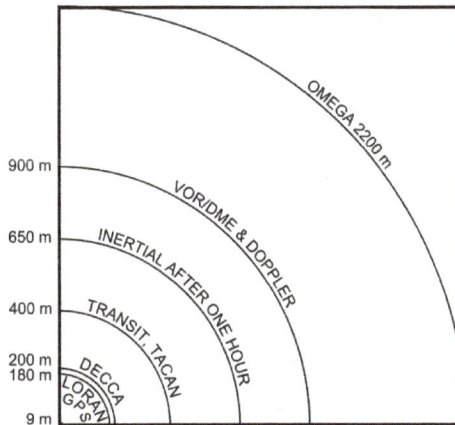

ACCURACY OF NAVIGATION SYSTEMS
(2-dimensional)

Radio navigation or radionavigation is the application of radio frequencies to determine a position of an object on the Earth. Like radiolocation, it is a type of radiodetermination.

- directions, e.g. by bearing, radio phases or interferometry,
- distances, e.g. ranging by measurement of travel times,
- partly also velocity, e.g. by means of radio Doppler shift.

Bearing-measurement Systems

These systems used some form of directional radio antenna to determine the location of a broadcast station on the ground. Conventional navigation techniques are then used to take a radio fix. These were introduced prior to WWI, and remain in use today.

Radio Direction Finding

The first system of radio navigation was the Radio Direction Finder, or RDF. By tuning in a radio station and then using a directional antenna, one could determine the direction to the broadcasting antenna. A second measurement using another station was then taken. Using triangulation, the two directions can be plotted on a map where their intersection reveals the location of the navigator. Commercial AM radio stations can be used for this task due to their long range and high power, but strings of low-power radio beacons were also set up specifically for this task, especially near airports and harbours.

Amelia Earhart's Lockheed Electra had a prominent RDF loop on the cockpit roof.

Early RDF systems normally used a loop antenna, a small loop of metal wire that is mounted so it can be rotated around a vertical axis. At most angles the loop has a fairly flat reception pattern, but when it is aligned perpendicular to the station the signal received on one side of the loop cancels the signal in the other, producing a sharp drop in reception known as the "null". By rotating the loop and looking for the angle of the null, the relative bearing of the station can be determined. Loop antennas can be seen on most pre-1950s aircraft and ships.

Reverse RDF

The Orfordness Beacon as it appears today.

The main problem with RDF is that it required a special antenna on the vehicle, which may not be easy to mount on smaller vehicles or single-crew aircraft. A smaller problem is that the accuracy of the system is based to a degree on the size of the antenna, but larger antennas would likewise make the installation more difficult.

During the era between World War I and World War II, a number of systems were introduced that placed the rotating antenna on the ground. As the antenna rotated through a fixed position, typically due north, the antenna was keyed with the morse code signal of the station's identification letters so the receiver could ensure they were listening to the right station. Then they waited for the signal to either peak or disappear as the antenna briefly pointed in their direction. By timing the delay between the morse signal and the peak/null, then dividing by the known rotational rate of the station, the bearing of the station could be calculated.

The first such system was the German Telefunken Kompass Sender, which began operations in 1907 and was used operationally by the Zeppelin fleet until 1918. An improved version was introduced by the UK as the Orfordness Beacon in 1929 and used until the mid-1930s. A number of improved versions followed, replacing the mechanical motion of the antennas with phasing techniques that produced the same output pattern with no moving parts. One of the longest lasting examples was Sonne, which went into operation just before World War II and used operationally under the name Consol until 1991. The modern VOR system is based on the same principles.

ADF and NDB

A great advance in the RDF technique was introduced in the form of phase comparisons of a signal as measured on two or more small antennas, or a single highly directional solenoid. These receivers were dramatically smaller, more accurate, and simpler to operate. Combined with the introduction of the transistor and integrated circuit, RDF systems were so reduced in size and complexity that they once again became quite common during the 1960s, and were known by the new name, automatic direction finder, or ADF.

This also led to a revival in the operation of simple radio beacons for use with these RDF systems, now referred to as non-directional beacons (NDB). As the LF/MF signals used by NDBs can follow the curvature of earth, NDB has a much greater range than VOR which travels only in line of sight. NDB can be categorized as long range or short range depending on their power. The frequency band allotted to non-directional beacons is 190–1750 kHz, but the same system can be used with any common AM-band commercial station.

VOR

VHF omnidirectional range, or VOR, is an implementation of the reverse-RDF system, but one that is more accurate and able to be completely automated.

Instead of a single signal, the VOR transmitter sends out three signals – one is a simple voice channel that sends morse code to identify the station, another is a continuous signal sent in all directions, and the last is a signal that is rotated at 30 RPM. Like the Orfordness concept, the bearing of the station is measured by finding the rotating signal's peak or null. But instead of timing the signal, the rotating signal is changed in phase in synchronicity with its rotation, such that it is in-phase when pointed north, 90 degrees off when it points east, and so forth. By comparing the phase of the received signal with the one being broadcast omnidirectionally, the angle can be determined using simple electronics. This angle is then displayed in the cockpit of the aircraft, and can be used to take a fix just like the earlier RDF systems, although it is easier to use.

VOR transmitter station

As VOR required two VHF receivers as well as a conventional radio for station identification, the system did not become popular until the era of miniaturized electronics, first with small tubes in the 1950s, and then transistorized systems in the 1960s. During this period it quickly took over from the older Radio Range system. The signals from the stations could be received anywhere, as opposed to the beams which were only broadcast in certain directions, so in theory the VOR system could be used for free navigation from any to any point. In practice, the older Radio Range procedures were so widely used and standardized that VOR was used to produce a similar set of airways that remain in use today.

The US military also introduced a VOR-like system known as TACAN. It differed from VOR primarily in its modulation system, adding a Lorentz-like signal to accurately define the center of the rotating beam and thereby improve accuracy. It requires five receiver channels and additional electronics, an expensive requirement when it was introduced.

Beam Systems

Beam systems broadcast narrow signals in the sky, and navigation is accomplished by keeping the aircraft centred in the beam. A number of stations are used to create an airway, with the navigator tuning in different stations along the direction of travel. These systems were common in the era when electronics were large and expensive, as they placed minimum requirements on the receivers – they were simply voice radio sets tuned to the selected frequencies. However, they did not provide navigation outside of the beams, and were thus less flexible in use. The rapid miniaturization of electronics during and after WWII made systems like VOR practical, and most beam systems rapidly disappeared.

Lorenz

In the post-WWI era, the Lorenz company of Germany developed a means of projecting two narrow radio signals with a slight overlap in the center. By broadcasting different audio signals in the two beams, the receiver could position themselves very accurately down the centreline by listening to the signal in their headphones. The system was accurate to less than a degree in some forms.

Originally known as "Ultrakurzwellen-Landefunkfeuer" (LFF), or simply "Leitstrahl" (guiding beam), little money was available to develop a network of stations. Deployment was instead led by the US, where it formed the basis of a wide-area navigation system through the 1930s and 40s. Development was restarted in Germany in the 1930s as a short-range system deployed at airports as a blind landing aid. Although there was some interest in deploying a medium-range system like the US LFF, deployment had not yet started when the beam system was combined with the Orfordness timing concepts to produce the highly accurate Sonne system. In all of these roles, the system was generically known simply as a "Lorenz beam".

In the immediate pre-WWII era the same concept was also developed as a blind-bombing system. This used very large antennas to provide the required accuracy at long distances (over England), and very powerful transmitters. Two such beams were used, crossing over the target to triangulate it. Bombers would enter one of the beams and use it for guidance until they heard the second one in a second radio receiver, using that signal to time the dropping of their bombs. The system was highly accurate, and the 'Battle of the Beams' broke out when United Kingdom intelligence services attempted, and then succeeded, in rendering the system useless through electronic warfare. Sonne, however, proved just as useful to the UK as Germany, and was left to operate unhindered throughout the war.

Low Frequency Radio Range

LFR ground station

The low-frequency radio range (LFR, also other names) was the main navigation system used by aircraft for instrument flying in the 1930s and 1940s in the U.S. and other countries, until the advent of the VOR in the late 1940s. It was used for both en route navigation as well as instrument approaches.

The ground stations consisted of a set of four antennas that projected Lorenz beams in four cardinal directions. One of the beams was "keyed" with the morse code signal "A", dit-dah, with the second beam "N", dah-dit. Flying down the centreline produced a steady tone. The beams were pointed to the next station to produce a set of airways, allowing an aircraft to travel from airport to airport by following a selected set of stations. Effective course accuracy was about three degrees, which near the station provided sufficient safety margins for instrument approaches down to low minimums. At its peak deployment, there were nearly 400 LFR stations in the US.

Glide Path and the Localizer of ILS

The remaining widely used beam systems are glide path and the localizer of the instrument landing system (ILS). ILS uses a localizer to provide horizontal position, distance to the runway, and airport information, and glide path to provide vertical positioning. ILS can provide enough accuracy and redundancy to allow automated landings.

Transponder Systems

Positions can be determined with any two measures of angle or distance. The introduction of radar in the 1930s provided a way to directly determine the distance to an object even at long distances. Navigation systems based on these concepts soon appeared, and remained in widespread use until recently. Today they are used primarily for aviation, although GPS has largely supplanted this role.

Radar and Transponders

Early radar systems, like the UK's Chain Home, consisted of large transmitters and separate receivers. The transmitter periodically sends out a short pulse of a powerful radio signal, which is sent into space through broadcast antennas. When the signal reflects off a target, some of that signal is reflected back in the direction of the station, where it is received. The received signal is a tiny fraction of the broadcast power, and has to be powerfully amplified in order to be used.

The same signals are also sent over local electrical wiring to the operator's station, which is equipped with an oscilloscope. Electronics attached to the oscilloscope provides a signal that increases in voltage over a short period of time, a few microseconds. When sent to the X input of the oscilloscope, this causes a horizontal line to be displayed on the scope. This "sweep" is triggered by a signal tapped off the broadcaster, so the sweep begins when the pulse is sent. Amplified signals from the receiver are then sent to the Y input, where any received reflection causes the beam to move upward on the display. This causes a series of "blips" to appear along the horizontal axis, indicating reflected signals. By measuring the distance from the start of the sweep to the blip, which corresponds to the time between broadcast and reception, the distance to the object can be determined.

Soon after the introduction of radar, the radio transponder appeared. Transponders are a combination of receiver and transmitter whose operation is automated – upon reception of a particular signal, normally a pulse on a particular frequency, the transponder sends out a pulse in response, typically delayed by some very short time. Transponders were initially used as the basis for early IFF systems; aircraft with the proper transponder would appear on the display as part of the normal radar operation, but then the signal from the transponder would cause a second blip to appear a short time later. Single blips were enemies, double blips friendly.

Transponder-based distance-distance navigation systems have a significant advantage in terms of positional accuracy. Any radio signal spreads out over distance, forming the fan-like beams of the Lorenz signal, for instance. As the distance between the broadcaster and receiver grows, the area covered by the fan increases, decreasing the accuracy of location within it. In comparison, transponder-based systems measure the timing between two signals, and the accuracy of that measure is largely a function of the equipment and nothing else. This allows these systems to remain accurate over very long range.

The latest transponder systems (mode S) can also provide position information, possibly derived from GNSS, allowing for even more precise positioning of targets.

Bombing Systems

The first distance-based navigation system was the German Y-Gerät blind-bombing system. This used a Lorenz beam for horizontal positioning, and a transponder for ranging. A ground-based system periodically sent out pulses which the airborne transponder returned. By measuring the total round-trip time on a radar's oscilloscope, the aircraft's range could be accurately determined even at very long ranges. An operator then relayed this information to the bomber crew over voice channels, and indicated when to drop the bombs.

The British introduced similar systems, notably the Oboe system. This used two stations in England that operated on different frequencies and allowed the aircraft to be triangulated in space. To ease pilot workload only one of these was used for navigation – prior to the mission a circle was drawn over the target from one of the stations, and the aircraft was directed to fly along this circle on instructions from the ground operator. The second station was used, as in Y-Gerät, to time the bomb drop. Unlike Y-Gerät, Oboe was deliberately built to offer very high accuracy, as good as 35 m, much better than even the best optical bombsights.

One problem with Oboe was that it allowed only one aircraft to be guided at a time. This was addressed in the later Gee-H system by placing the transponder on the ground and broadcaster in the aircraft. The signals were then examined on existing Gee display units in the aircraft. Gee-H did not offer the accuracy of Oboe, but could be used by as many as 90 aircraft at once. This basic concept has formed the basis of most distance measuring navigation systems to this day.

Beacons

The key to the transponder concept is that it can be used with existing radar systems. The ASV radar introduced by RAF Coastal Command was designed to track down submarines and ships by displaying the signal from two antennas side by side and allowing the operator to compare their relative strength. Adding a ground-based transponder immediately turned the same display into a system able to guide the aircraft towards a transponder, or "beacon" in this role, with high accuracy.

The British put this concept to use in their Rebecca/Eureka system, where battery-powered "Eureka" transponders were triggered by airborne "Rebecca" radios and then displayed on ASV Mk. II radar sets. Eureka's were provided to French resistance fighters, who used them to call in supply drops with high accuracy. The US quickly adopted the system for paratroop operations, dropping the Eureka with pathfinder forces or partisans, and then homing in on those signals to mark the drop zones.

The beacon system was widely used in the post-war era for blind bombing systems. Of particular note were systems used by the US Marines that allowed the signal to be delayed in such a way to offset the drop point. These systems allowed the troops at the front line to direct the aircraft to points in front of them, directing fire on the enemy. Beacons were widely used for temporary or mobile navigation as well, as the transponder systems were generally small and low-powered, able to be man portable or mounted on a Jeep.

DME

In the post-war era, a general navigation system using transponder-based systems was deployed as the distance measuring equipment (DME) system.

DME was identical to Gee-H in concept, but used new electronics to automatically measure the time delay and display it as a number, rather than having the operator time the signals manually on an oscilloscope. This led to the possibility that DME interrogation pulses from different aircraft might be confused, but this was solved by having each aircraft send out a different series of pulses which the ground-based transponder repeated back.

DME is almost always used in conjunction with VOR, and is normally co-located at a VOR station. This combination allows a single VOR/DME station to provide both angle and distance, and thereby provide a single-station fix. DME is also used as the distance-measuring basis for the military TACAN system, and their DME signals can be used by civilian receivers.

Hyperbolic Systems

Hyperbolic navigation systems are a modified form of transponder systems which eliminate the need for an airborne transponder. The name refers to the fact that they do not produce a single distance or angle, but instead indicate a location along any number of hyperbolic lines in space. Two such measurements produces a fix. As these systems are almost always used with a specific navigational chart with the hyperbolic lines plotted on it, they generally reveal the receiver's location directly, eliminating the need for manual triangulation. As these charts were digitized, they became the first true location-indication navigational systems, outputting the location of the receiver as latitude and longitude. Hyperbolic systems were introduced during WWII and remained the main long-range advanced navigation systems until GPS replaced them in the 1990s.

Gee

The first hyperbolic system to be developed was the British Gee system, developed during World War II. Gee used a series of transmitters sending out precisely timed signals, with the signals leaving the stations at fixed delays. An aircraft using Gee, RAF Bomber Command's heavy bombers, examined the time of arrival on an oscilloscope at the navigator's station. If the signal from two stations arrived at the same time, the aircraft must be an equal distance from both transmitters, allowing the navigator to determine a line of position on his chart of all the positions at that distance from both stations. More typically, the signal from one station would be received earlier than the other. The difference in timing between the two signals would reveal them to be along a curve of possible locations. By making similar measurements with other stations, additional lines of position can be produced, leading to a fix. Gee was accurate to about 165 yards (150 m) at short ranges, and up to a mile (1.6 km) at longer ranges over Germany. Gee remained in use long after WWII, and equipped RAF aircraft as late as the 1960s (approx freq was by then 68 MHz).

LORAN

With Gee entering operation in 1942, similar US efforts were seen to be superfluous. They turned their development efforts towards a much longer-ranged system based on the same principles, using much lower frequencies that allowed coverage across the Atlantic Ocean. The result was LORAN, for "LOng-range Aid to Navigation". The downside to the long-wavelength approach was that accuracy was greatly reduced compared to the high-frequency Gee. LORAN was widely used during convoy operations in the late war period.

Decca

Another British system from the same era was Decca Navigator. This differed from Gee primarily in that the signals were not pulses delayed in time, but continuous signals delayed in phase. By comparing the phase of the two signals, the time difference information as Gee was returned. However, this was far easier to display; the system could output the phase angle to a pointer on a dial removing any need for visual interpretation. As the circuitry for driving this display was quite small, Decca systems normally used three such displays, allowing quick and accurate reading of multiple fixes. Decca found its greatest use post-war on ships, and remained in use into the 1990s.

LORAN-C

Almost immediately after the introduction of LORAN, in 1952 work started on a greatly improved version. LORAN-C (the original retroactively became LORAN-A) combined the techniques of pulse timing in Gee with the phase comparison of Decca.

The resulting system (operating in the low frequency (LF) radio spectrum from 90 to 110 kHz) that was both long-ranged (for 60 kW stations, up to 3400 miles) and accurate. To do this, LORAN-C sent a pulsed signal, but modulated the pulses with an AM signal within it. Gross positioning was determined using the same methods as Gee, locating the receiver within a wide area. Finer accuracy was then provided by measuring the phase difference of the signals, overlaying that second measure on the first. By 1962, high-power LORAN-C was in place in at least 15 countries.

LORAN-C was fairly complex to use, requiring a room of equipment to pull out the different signals. However, with the introduction of integrated circuits, this was quickly reduced further and further. By the late 1970s LORAN-C units were the size of a stereo amplifier and were commonly found on almost all commercial ships as well as some larger aircraft. By the 1980s this had been further reduced to the size of a conventional radio, and it became common even on pleasure boats and personal aircraft. It was the most popular navigation system in use through the 1980s and 90s, and its popularity led to many older systems being shut down, like Gee and Decca. However, like the beam systems before it, civilian use of LORAN-C was short-lived when GPS technology drove it from the market.

Other Hyperbolic Systems

Similar hyperbolic systems included the US global-wide VLF/Omega Navigation System, and the similar Alpha deployed by the USSR. These systems determined pulse timing not by comparison of two signals, but by comparison of a single signal with a local atomic clock. The expensive-to-maintain Omega system was shut down in 1997 as the US military migrated to using GPS. Alpha is still in use.

Satellite Navigation

Since the 1960s, navigation has increasingly moved to satellite navigation systems. These are essentially DME systems located in space. The fact that the satellites are in orbit and normally move with respect to the receiver means that the calculation of the position of the satellite needs to be taken into account as well, which can only be handled effectively with a computer.

Cessna 182 with GPS-based "glass cockpit" avionics

The Global Positioning System, better known simply as GPS, sends several signals that are used to decode the position and distance of the satellite. One signal encodes the satellite's "ephemeris" data, which is used to accurately calculate the satellite's location at any time. Space weather and other effects causes the orbit to change over time so the ephemeris has to be updated periodically. Other signals send out the time as measured by the satellite's onboard atomic clock. By measuring this signal from several satellites, the receiver can rebuild an accurate clock signal of its own. Comparing the two produces the distance to the satellite, and several such measurements allows a form of triangulation to be carried out.

GPS has better accuracy that any previous land-based system, is available at almost all locations on the Earth, can be implemented in a few cents of modern electronics, and requires only a few dozen satellites to provide worldwide coverage. As a result of these advantages, GPS has led to almost all previous systems falling from use. LORAN, Omega, Decca, Consol and many other systems disappeared during the 1990s and 2000s. The only other systems still in use are aviation aids, which are also being turned off for long-range navigation while new differential GPS systems are being deployed to provide the local accuracy needed for blind landings.

Radiolocation

Radiolocating is the process of finding the location of something through the use of radio waves. It generally refers to passive uses, particularly radar—as well as detecting buried cables, water mains, and other public utilities. It is similar to radionavigation, but radiolocation usually refers to passively finding a distant object rather than actively one's own position. Both are types of radiodetermination. Radiolocation is also used in real-time locating systems (RTLS) for tracking valuable assets.

Basic Principles

An object can be located by measuring the characteristics of received radio waves. The radio waves may be transmitted by the object to be located, or they may be backscattered waves (as in radar or passive RFID). A stud finder uses radiolocation when it uses radio waves rather than ultrasound.

One technique measures a distance by using the difference in the power of the received signal strength (RSSI) as compared to the originating signal strength. Another technique uses the time of arrival (TOA), when the time of transmission and speed of propagation are known. Combining TOA data from several receivers at different known locations (time difference of arrival, TDOA) can provide an estimate of position even in the absence of knowledge of the time of transmission. The angle of arrival (AOA) at a receiving station can be determined by the use of a directional antenna, or by differential time of arrival at an array of antennas with known location. AOA information may be combined with distance estimates from the techniques previously described to establish the location of a transmitter or backscatterer. Alternatively, the AOA at two receiving stations of known location establishes the position of the transmitter. The use of multiple receivers to locate a transmitter is known as multilateration.

Estimates are improved when the transmission characteristics of the medium is factored into the calculations. For RSSI this means electromagnetic permeability; for TOA it may mean non-line-of-sight receptions.

Use of RSSI to locate a transmitter from a single receiver requires that both the transmitted (or backscattered) power from the object to be located are known, and that the propagation characteristics of the intervening region are known. In empty space, signal strength decreases as the inverse square of the distance for distances large compared to a wavelength and compared to the object to be located, but in most real environments, a number of impairments can occur: absorption, refraction, shadowing, and reflection. Absorption is negligible for radio propagation in air at frequencies less than about 10 GHz, but becomes important at multi-GHz frequencies where rotational molecular states can be excited. Refraction is important at long ranges (tens to hundreds of kilometers) due to gradients in moisture content and temperature in the atmosphere. In urban, mountainous, or indoor environments, obstruction by intervening obstacles and reflection from nearby surfaces are very common, and contribute to multipath distortion: that is, reflected and delayed replicates of the transmitted signal are combined at the receiver. Signals from different paths can add constructively or destructively:

such variations in amplitude are known as fading. The dependence of signal strength on position of transmitter and receiver becomes complex and often non-monotonic, making single-receiver estimates of position inaccurate and unreliable. Multilateration using many receivers is often combined with calibration measurements ("fingerprinting") to improve accuracy.

TOA and AOA measurements are also subject to multipath errors, particularly when the direct path from the transmitter to receiver is blocked by an obstacle. Time of arrival measurements are also most accurate when the signal has distinct time-dependent features on the scale of interest—for example, when it is composed of short pulses of known duration—but Fourier transform theory shows that in order to change amplitude or phase on a short time scale, a signal must use a broad bandwidth. For example, to create a pulse of about 1 ns duration, roughly sufficient to identify location to within 0.3 m (1 foot), a bandwidth of roughly 1 GHz is required. In many regions of the radio spectrum, emission over such a broad bandwidth is not allowed by the relevant regulatory authorities, in order to avoid interference with other narrowband users of the spectrum. In the United States, unlicensed transmission is allowed in several bands, such as the 902-928 MHz and 2.4-2.483 GHz Industrial, Scientific, and Medical ISM bands, but high-power transmission cannot extend outside of these bands. However, several jurisdictions now allow ultrawideband transmission over GHz or multi-GHz bandwidths, with constraints on transmitted power to minimize interference with other spectrum users. UWB pulses can be very narrow in time, and often provide accurate estimates of TOA in urban or indoor environments.

Radiolocation is employed in a wide variety of industrial and military activities. Radar systems often use a combination of TOA and AOA to determine a backscattering object's position using a single receiver. In Doppler radar, the Doppler shift is also taken into account, determining velocity rather than location (though it helps determine future location). Real Time Location Systems RTLS using calibrated RTLS, and TDOA, are commercially available. The widely used Global Positioning System (GPS) is based on TOA of signals from satellites at known positions.

Mobile Phones

Radiolocation is also used in cellular telephony via base stations. Most often, this is done through trilateration between radio towers. The location of the Caller or handset can be determined several ways:

- angle of arrival (AOA) requires at least two towers, locating the caller at the point where the lines along the angles from each tower intersect

- time difference of arrival (TDOA) resp. time of arrival (TOA) works using multilateration, except that it is the networks that determine the time difference and therefore distance from each tower (as with seismometers)

- location signature uses "fingerprinting" to store and recall patterns (such as multipath) which mobile phone signals are known to exhibit at different locations in each cell

The first two depend on a line-of-sight, which can be difficult or impossible in mountainous terrain or around skyscrapers. Location signatures actually work better in these conditions however. TDMA and GSM networks such as Cingular and T-Mobile use TDOA.

CDMA networks such as Verizon Wireless and Sprint PCS tend to use handset-based radiolocation technologies, which are technically more similar to radionavigation. GPS is one of those technologies.

Composite solutions, needing both the handset and the network include:

- assisted GPS (wireless or TV) allows use of GPS even indoors

- Advanced Forward Link Trilateration (A-FLT)

- Timing Advance/Network Measurement Report (TA/NMR)

- Enhanced Observed Time Difference (E-OTD)

Initially, the purpose of any of these in mobile phones is so that the public safety answering point (PSAP) which answers calls to an emergency telephone number can know where the caller is and exactly where to send emergency services. This ability is known within the NANP (North America) as wireless enhanced 911. Mobile phone users may have the option to permit the location information gathered to be sent to other phone numbers or data networks, so that it can help people who are simply lost or want other location-based services. By default, this selection is usually turned off, to protect privacy.

Direction Finding

Direction finding (DF), or radio direction finding (RDF), is the measurement of the direction from which a received signal was transmitted. This can refer to radio or other forms of wireless communication, including radar signals detection and monitoring (ELINT/ESM). By combining the direction information from two or more suitably spaced receivers (or a single mobile receiver), the source of a transmission may be located via triangulation. Radio direction finding is used in the navigation of ships and aircraft, to locate emergency transmitters for search and rescue, for tracking wildlife, and to locate illegal or interfering transmitters. RDF was important in combating German threats during both the WW-II Battle of Britain and the long running Battle of the Atlantic. In the former, the Air Ministry also used RDF to locate its own fighter groups and vector them to detected German raids.

RDF systems can be used with any radio source, although very long wavelengths (low frequencies) require very large antennas, and are generally used only on ground-based systems. These wavelengths are nevertheless used for marine radio navigation as they can travel very long distances "over the horizon", which is valuable for ships when the line-of-sight may be only a few tens of kilometres. For aerial use, where the horizon may extend to hundreds of kilometres, higher frequencies can be used, allowing the use of much smaller antennas. An automatic direction finder, which could be tuned to radio beacons called non-directional beacons or commercial AM radio broadcasters, was until recently, a feature of most aircraft, but is now being phased out

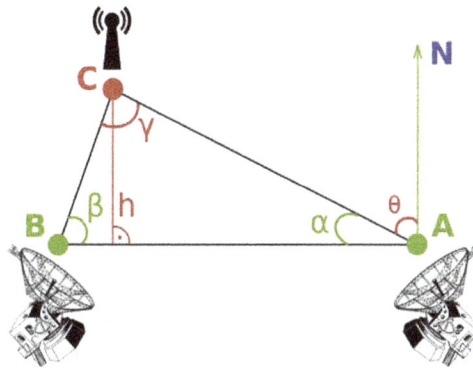

Radiotriangulation scheme

For the military, RDF is a key tool of signals intelligence. The ability to locate the position of an enemy transmitter has been invaluable since World War I, and played a key role in World War II's Battle of the Atlantic. It is estimated that the UK's advanced "huff-duff" systems were directly or indirectly responsible for 24% of all U-Boats sunk during the war. Modern systems often used phased array antennas to allow rapid beamforming for highly accurate results, and are part of a larger electronic warfare suite.

Radio direction finders have evolved, following the development of new electronics. Early systems used mechanically rotated antennas that compared signal strengths, and several electronic versions of the same concept followed. Modern systems use the comparison of phase or doppler techniques which are generally simpler to automate. Early British radar sets were referred to as RDF, which is often stated was a deception. In fact, the Chain Home systems used large RDF receivers to determine directions. Later radar systems generally used a single antenna for broadcast and reception, and determined direction from the direction the antenna was facing.

Antennas

Direction finding requires an antenna that is directional (more sensitive in certain directions than in others). Many antenna designs exhibit this property. For example, a Yagi antenna has quite pronounced directionality, so the source of a transmission can be determined simply by pointing it in the direction where the maximum signal level is obtained. However, to establish direction to great accuracy requires more sophisticated technique.

A simple form of directional antenna is the loop aerial. This consists of an open loop of wire on an insulating former, or a metal ring that forms the antenna elements itself, where the diameter of the loop is a tenth of a wavelength or smaller at the target frequency. Such an antenna will be least sensitive to signals that are normal to its face and most responsive to those meeting edge-on. This is caused by the phase output of the transmitting beacon. The phase changing phase causes a difference between the voltages induced on either side of the loop at any instant. Turning the loop face on will not induce any current flow. Simply turning the antenna to obtain minimum signal will establish two possible directions from which the signal could be emanating. The NULL is used, as small angular deflections of the loop aerial near its null positions produce larger changes in current than similar angular changes near the loops max positions. For this reason, a null position of the loop aerial is used.

The crossed-loops DF antenna atop the mast of a tug boat

To resolve the two direction possibilities, a sense antenna is used, the sense aerial has no directional properties but has the same sensitivity as the loop aerial. By adding the steady signal from the sense aerial to the alternating signal from the loop signal as it rotates, there is now only one position as the loop rotates 360° at which there is zero current. This acts as a phase ref point, allowing the correct null point to be identified, thus removing the 180° ambiguity. A dipole antenna exhibits similar properties, and is the basis for the Yagi antenna, which is familiar as the common VHF or UHF television aerial. For much higher frequencies still, parabolic antennas can be used, which are highly directional, focusing received signals from a very narrow angle to a receiving element at the centre.

More sophisticated techniques such as phased arrays are generally used for highly accurate direction finding systems called goniometers such as are used in signals intelligence (SIGINT). A helicopter based DF system was designed by ESL Incorporated for the U.S. Government as early as 1972.

The RDF antenna on this B-17F is located in the prominent teardrop housing under the nose.

Single Channel DF or Single Site Location (SSL)

Single-channel DF uses a multi-antenna array with a single channel radio receiver. This approach to DF obviously offers some advantages and drawbacks. Since it only uses one receiver, mobility and lower power consumption are obvious benefits but without the ability to look at each antenna simultaneously (which would be the case if one were to use multiple receivers, also known as N-channel DF) more complex operations need to occur at the antenna in order to present the signal to the receiver.

The two main categories that a single channel DF algorithm falls into are amplitude comparison and phase comparison. Some algorithms can be hybrids of the two.

Pseudo-doppler DF Technique

The pseudo-doppler technique is a phase based DF method that produces a bearing estimate on the received signal by measuring the doppler shift induced on the signal by sampling around the elements of a circular array. The original method used a single antenna that physically moved in a circle but the modern approach uses a multi-antenna circular array with each antenna sampled in succession.

Watson-Watt/Adcock Antenna Array

The Watson-Watt technique uses two Adcock antenna pairs to perform an amplitude comparison on the incoming signal. An Adcock antenna pair is a pair of monopole or dipole antennas that takes the vector difference of the received signal at each antenna so that there is only one output from the pair of antennas. Two of these pairs are co-located but perpendicularly oriented to produce what can be referred to as the N-S (North-South) and E-W (East-West) signals that will then be passed to the receiver. In the receiver, the bearing angle can then be computed by taking the arctangent of the ratio of the N-S to E-W signal.

Correlative Interferometer

The basic principle of the correlative interferometer consists in comparing the measured phase differences with the phase differences obtained for a DF antenna system of known configuration at a known wave angle (reference data set). The comparison is made for different azimuth values of the reference data set, the bearing is obtained from the data for which the correlation coefficient is at a maximum. If the direction finding antenna elements have a directional antenna pattern, then the amplitude may be included in the comparison.

Usage

Radio Navigation

Radio direction finding, Radio direction finder, or RDF was once the primary aviation navigational aid. (Range and Direction Finding was the abbreviation used to describe the predecessor to Radar.) Beacons were used to mark "airways" intersections and to define departure and approach procedures. Since the signal transmitted contains no information about bearing or distance, these beacons are referred to as non-directional beacons, or NDB in the aviation world. Starting in the 1950s, these beacons were generally replaced by the VOR system, in which the bearing to the navigational aid is measured from the signal itself; therefore no specialized antenna with moving parts is required. Due to relatively low purchase, maintenance and calibration cost, NDB's are still used to mark locations of smaller aerodromes and important helicopter landing sites.

A portable, battery operated GT-302 Accumatic automatic direction finder for marine use

Similar beacons located in coastal areas are also used for maritime radio navigation, as almost every ship is (was) equipped with a direction finder (Appleyard 1988). Very few maritime radio navigation beacons remain active today (2008) as ships have abandoned navigation via RDF in favor of GPS navigation.

In the United Kingdom a radio direction finding service is available on 121.5 MHz and 243.0 MHz to aircraft pilots who are in distress or are experiencing difficulties. The service is based on a number of radio DF units located at civil and military airports and certain HM Coastguard stations. These stations can obtain a "fix" of the aircraft and transmit it by radio to the pilot.

Location of Illegal, Secret or Hostile Transmitters - SIGINT

In WW2 considerable effort was expended on identifying secret transmitters in the United Kingdom (UK) by direction finding. The work was undertaken by the Radio Security Service (RSS also MI8). Initially three U Adcock HF DF stations were set up in 1939 by the General Post Office. With the declaration of war, MI5 and RSS developed this into a larger network. One of the problems with providing coverage of an area the size of the UK was installing sufficient DF stations to cover the entire area to receive skywave signals reflected back from the ionised layers in the upper atmosphere. Even with the expanded network, some areas were not adequately covered and for this reason up to 1700 voluntary interceptors (radio amateurs) were recruited to detect illicit transmissions by ground wave. In addition to the fixed stations, RSS ran a fleet of mobile DF vehicles around the UK. If a transmitter was identified by the fixed DF stations or voluntary interceptors, the mobile units were sent to the area to home in on the source. The mobile units were HF Adcock systems.

British Post Office RDF lorry from 1927 for finding unlicensed amateur radio transmitters. It was also used to find regenerative receivers which radiated interfering signals due to feedback, a big problem at the time.

By 1941 only a couple of illicit transmitters had been identified in the UK; these were German agents that had been "turned" and were transmitting under MI5 control. Many illicit transmissions had been logged emanating from German agents in occupied and neutral countries in Europe. The traffic became a valuable source of intelligence, so the control of RSS was subsequently passed to MI6 who were responsible for secret intelligence originating from outside the UK. The direction finding and interception operation increased in volume and importance until 1945.

The HF Adcock stations consisted of four 10m vertical antennas surrounding a small wooden operators hut containing a receiver and a radio-goniometer which was adjusted to obtain the bearing. MF stations were also used which used four guyed 30m lattice tower antennas. In 1941 RSS began experimenting with Spaced Loop direction finders, developed by the Marconi company and the UK National Physical Laboratories. These consisted of two parallel loops 1 to 2m square on the ends of a rotatable 3 to 8m beam. The angle of the beam was combined with results from a radiogoniometer to provide a bearing. The bearing obtained was considerably sharper than that obtained with the U Adcock system, but there were ambiguities which prevented the installation of 7 proposed S.L DF systems. The operator of an SL system was in a metal underground tank below the antennas. Seven underground tanks were installed, but only two SL systems were installed at Wymondham, Norfolk and Weaverthorp in Yorkshire. Problems were encountered resulting in the remaining five underground tanks being fitted with Adcock systems. The rotating SL antenna was turned by hand which meant successive measurements were a lot slower than turning the dial of a goniometer.

Another experimental spaced loop station was built near Aberdeen in 1942 for the Air Ministry with a semi-underground concrete bunker. This, too, was abandoned because of operating difficulties. By 1944 a mobile version of the spaced loop had been developed and was used by RSS in France following the D-Day invasion of Normandy.

The US military used a shore based version of the spaced loop DF in WW2 called "DAB". The loops were placed at the ends of a beam, all of which was located inside a wooden hut with the electronics in a large cabinet with cathode ray tube display at the centre of the beam and everything being supported on a central axis. The beam was rotated manually by the operator.

The Royal Navy introduced a variation on the shore based HF DF stations in 1944 to track U-boats in the North Atlantic. They built groups of five DF stations, so that bearings from individual stations in the group could be combined and a mean taken. Four such groups were built in Britain at Ford End, Essex, Goonhavern, Cornwall, Anstruther and Bowermadden in the Scottish Highlands. Groups were also built in Iceland, Nova Scotia and Jamaica. The anticipated improvements were not realised but later statistical work improved the system and the Goonhavern and Ford End groups continued to be used during the Cold War. The Royal Navy also deployed direction finding equipment on ships tasked to anti-submarine warfare in order to try to locate German submarines, e.g. Captain class frigates were fitted with a medium frequency direction finding antenna (MF/DF) (the antenna was fitted in front of the bridge) and high frequency direction finding (HF/DF, "Huffduff") Type FH 4 antenna (the antenna was fitted on top of the mainmast).

Arguably the most comprehensive reference on WW2 wireless direction finding was written by Roland Keen who was head of the engineering department of RSS at Hanslope Park. The DF systems mentioned here are described in detail in his exhaustive treatment of the subject in the 1947 edition of his book "Wireless Direction Finding".

At the end of WW2 a number of RSS DF stations continued to operate into the cold war under the control of GCHQ the British SIGINT organisation.

Most direction finding effort within the UK now (2009) is directed towards locating unauthorised "pirate" FM broadcast radio transmissions. A network of remotely operated VHF direction finders are used mainly located around the major cities. The transmissions from mobile telephone handsets are also located by a form of direction finding using the comparative signal strength at the surrounding local "cell" receivers. This technique is often offered as evidence in UK criminal prosecutions and, almost certainly, for SIGINT purposes.

Emergency Aid

There are many forms of radio transmitters designed to transmit as a beacon in the event of an emergency, which are widely deployed on civil aircraft. Modern emergency beacons transmit a unique identification signal that can aid in finding the exact location of the transmitter.

Avalanche Rescue

Avalanche transceivers operate on a standard 457 kHz, and are designed to help locate people and equipment buried by avalanches. Since the power of the beacon is so low the directionality of the radio signal is dominated by small scale field effects and can be quite complicated to locate.

Wildlife Tracking

Location of radio-tagged animals by triangulation is a widely applied research technique for studying the movement of animals. The technique was first used in the early 1960s, when the technology used in radio transmitters and batteries made them small enough to attach to wild animals, and is now widely deployed for a variety of wildlife studies. Most tracking of wild animals that have been affixed with radio transmitter equipment is done by a field researcher using a handheld radio direction finding device. When the researcher wants to locate a particular animal, the location of the animal can be triangulated by determining the direction to the transmitter from several locations.

Reconnaissance

Phased arrays and other advanced antenna techniques are utilized to track launches of rocket systems and their resulting trajectories. These systems can be used for defensive purposes and also to gain intelligence on operation of missiles belonging to other nations. These same techniques are used for detection and tracking of conventional aircraft.

Sport

Events hosted by groups and organizations that involve the use of radio direction finding skills to locate transmitters at unknown locations have been popular since the end of World War II. Many of these events were first promoted in order to practice the use of radio direction finding techniques for disaster response and civil defense purposes, or to practice locating the source of radio frequency interference. The most popular form of the sport, worldwide, is known as Amateur Radio Direction Finding or by its international abbreviation ARDF. Another form of the activity, known as "transmitter hunting", "mobile T-hunting" or "fox hunting" takes place in a larger geographic area, such as the metropolitan area of a large city, and most participants travel in motor vehicles while attempting to locate one or more radio transmitters with radio direction finding techniques.

Fuzzy Locating System

Fuzzy locating is a rough but reliable method based on appropriate measuring technology for estimating a location of an object. The concept of precise or crisp locating is replaced with respect to the operational requirements and the economic viability. In most cases the knowledge of exact coordinates does not contribute to operations, but the spatial or planar relation between entities is relevant. Hence fuzzy locating determines the radial distances between entities involved in an operational process and reduces the required accuracy of measurement to basic qualities of close, near or far and to relations simple as in or out. However such segregation shall be achieved with high reliability and sound repetition.

Basics

The term fuzzy relates to rough spatial coincidence or contiguity assessment compared to the alternative crisp locating, which derives precise coordinates of a location of an object. The fuzzy part of the locating process balances the physical and the mathematical portions of processing measurement data of the objects involved and a priori knowledge with the operational ambience.

The result of fuzzy locating shall suffice for operational support and not for metric confirmation of measures taken at earlier occasions. However, available information is exploited as a priori knowledge. Fuzzy locating compares with the distinction respectively segregation of mathematical logic with the terms crisp sets and crisp relations or fuzzy sets and fuzzy relations.

Systematic and stochastic errors occurring under operational requirements and conditions turn virtually precise measures in a friendly ambience to fuzzy metrics. In even worse ambient conditions, which mostly applies to wireless propagation in ISM bands, this leads to erratic results and various misinterpretation. In consequence, the trade-off between technical effort and achieved operational support adjusts inevitably to physical limitations as well as to weaknesses in mathematical modeling. Better resolved balancing deliberately neglects classical terms of precision in favour of a strong commitment for operational unambiguity.

The Techno-economic Challenge

Generally, precision is obtained at expense. The balancing of capital expenditure and of operational cost shall take into account not what is possible, but what is necessary. A better designed balancing leads to the less precise fuzzy locating at much lesser expense: As a more general approach, fuzzy locating is a method for best estimating a roughly determined location of an object as a distinction of operational contiguity. Contiguity means more or less a handy distance between an individual and an object. Three basic situations generally apply:

- Basic task is, for example, current presence in a room, well discriminated from other adjacent rooms.

- The more challenging requirement is to segregate the presence of an object or an individual at just one of several work positions in the very same room or with any other known and well referred place in this room.

- The more generalized approach hence is the spatial relationship between an object and an individual in any ambience, defining the actual location of either the object or alternatively the individual as the point of reference.

In all three cases the absolute coordinates are not of interest, as long as the discrimination of rooms or work positions is reliably achieved.

Wireless, Optical and Acoustical Approaches

For the locating task, an object to be located must be at least equipped with a wireless tag in a wireless communication environment. Each operating wireless target in a wireless communication environment may contribute. Prerequisite for radio frequency based wireless cooperation is some cohesion of the wireless nodes in a networking concept. Each wireless target has at least one physical propagation parameter that varies with location. Better qualified approaches make use of more than one physical parameter.

Alternatively, optical and acoustical solutions are known. Variation of parameters is partly deterministic with varying distances between wireless nodes. A location estimate approximating the real location of the transmitter, preferably under real time constraints, is determined on the basis of a stochastical model of propagation and a model for the process of observation in a noisy ambience and on a chosen set of observed deterministic parameters of transmission and propagation.

Implemented Examples

Several suppliers offer the so-called electronic leash solution. This serves for wirelessly tethering mobile appliances with each other. The RSSI estimate serves for a radial metrics but without any certified calibration. Setting an alarm on unintentional loss is the key service offered with this concept. An advanced aspect has been launched with Bluetooth low energy for better economized battery life cycle. Special trimming serves for two years operation from a button cell.

Comparison to Metric Locating

Metric or crisp locating determines a spatial or planar relationship between independently moving or residing entities (usually addressed as targets) by means of qualified methods for measuring distances. This is the topic for example of

- satellite positioning systems, as GPS or Galileo

- real time locating systems (RTLS) as defined with ISO/IEC 19762-5 or

- inertial navigation systems (INS)

These technologies generally make use of a travel time measurement as the approach with best resolution and precision. Further enhancement is achieved with time differences discriminated for several paths. Such basic or enhanced travel time measurement requires a multiplicity of measures for unambiguous locating.

All of these sophisticated physical methods of measuring are hampered with a challenge caused by motion and caused by transmitter population. This makes restrictions effective in time both for observation and capture and for communication of measurement data. In consequence, the pecuniary and the technical effort adjusts to physical limitations and the limited metric precision with a special aspect to operational clarity. Such balancing neglects classical terms of metric precision, prevents from over-interpreting erratic measures and provides sufficient escape.

Additionally the model of propagation contributes to the achieved results. In satellite based systems, direct line of sight is generally required, without escape. That determines the restriction with applying such approach between buildings or, even worse, inside buildings: The highest precision does not compensate for bad visibility. Whenever the path of propagation gets cranked, the result of time measurement gets biased.

In comparison to locating on the move, exactly determining a location with highest possible precision is the topic of geodesy and surveying. These disciplines traditionally do not deal with motion and may integrate over long time. The terms of 'locating', 'positioning' and 'navigating' or 'surveying' are commonly used in almost equivalence, hence neglecting that the sense of these terms is different concerning sensor and actor functions and motion conditions.

Radio Signal Strength Indication (RSSI) as a Coarse Metric

For many purposes, a distinct and reliable determination of a location relative to somewhat rastered position on a floor or just a room in a building will be sufficient for sound fuzzy locating by single measurements.

Typically fuzzy locating coincides with simple power level measurement, usually configured as unilateration. A combination of multiple distance measurement, as a multilateration, based on power level measurement, appears unbalanced. The effort for multiple measures aiming at an unambiguous multilateration process will not be justified with the achievable precision of the results of power level measures.

Though leaving some ambiguity with sparse measurements, the contiguity may be assessed applying a priori knowledge. Such includes primarily tracking motion over time. As a generalized approach the fuzzy locating based on power levels measured with wireless nodes will roughly suffice for coarse guessing a location, where an object or a person with a wireless node resides in contiguity to other wireless nodes in a wireless environment.

Therefore each wireless node recognises the received power level with the distance to the transmitting wireless node and this parameter can be measured. Some additional calibration parameters serve as a basis for the statistical model of the parameters of propagation in a known neighborhood of distributed wireless nodes and of other passive objects, which influence propagation. A location estimate, which approximates the approach of the transmitting node, will be described with the stochastic model of propagation and the statistics of the monitored parameters.

Offered Discrimination of Positions

For applications where no need for absolute coordinates determination is assessed, the implementing of a more simple solution is advantageous. Compared to multilateration as the concept of crisp locating, the other option is fuzzy locating, where just one distance delivers the relation between detector and detected object. This most simple approach is unilateration. However, such unilateration approach never delivers the angular position with reference to the detector. Many solutions are available today.

Offered Qualities with Auxiliary Mapping

Increasing accuracy means increasing cost. The most indirect approach is the increase of distributed anchor nodes. The first direct approach simply is a fixed excitation through wall-mounted wireless nodes or optical exciters. That will provide a sound room discrimination in any case. The second direct approach is the position discrimination using apparently available infrastructure objects, as with networked work stations yet equipped with Bluetooth transponders.

The easy escape beyond increase of accuracy in measuring is the accuracy gain with mapping. Such mapping seldom suffices when performed statically. The more advantageous approach is a combination of initial mapping based on floor plans or area maps with an intelligent update based on updates obtained from actual measuring. That is offered with the concept of simultaneous Locating and Mapping.

Physical Restrictions

Measurement of propagation parameters is generally heavily loaded with various noise components. Such noise may be stochastic ([white noise]) or deterministic or a mixture of noises with limited bandwidth ([pink noise]).

As with all wireless systems, qualities of measurement in any one aspect contradict to qualities in some other aspects:

- Especially extension of reach collides with reducing the separation limits.

- Also increase of probing duration collides with motion speed.

- Generally the requirement for unambiguous locating requires a set of measured distances.

Such contracdictions result in challenges for systems performance and in hard restrictions concerning timely parallel availability of certain quality levels. Facing limitations as inevitable, the implementer of a locating systems has to determine the operational requirements first and then has to make a choice under a set of alternatives and must scale the adjoint limitations. The outcome is always a compromise with trade offs usually in budgetary effort and in technical precision. Any chosen alternative generally will exclude certain other technical options from operational availability.

Noisy Ambience as a General Condition

In technical terms, operational environments are generally noisy. For measuring, that is not a friendly ambience. Systematic and stochastic errors under operational requirements and conditions turn virtually precise measures to fuzzy metrics. The measuring results get in worse in a densely populated ambience, especially in the vicinity of other electrically active objects. Even physically passive surfaces contribute to the measurement problem. All physical effects tend to contribute unsteady and non linear behavior. Hence the physical measurement errors lead to biased or erratic results under such noisy conditions.

Frequency Spread as an Option

One escape from collision problems with wireless networks is the separation of the measuring and the communications processes with allocation to different frequencies, however requiring respective dual transponder capabilities. Another escape from collision problems with wireless networks is the spread of signals with stochastic coding.

Time as a Restriction for Measuring

Even if targets do not move, time is a restriction with the performing of a locating function. The first impact is that of allowed minimal distinct time differences that define the theoretically best resolution. This varies conceptually between phase discrimination in fractions of a cycle to be measured and full cycles to be counted. In competition for non-colliding transmission, time may appear as the main aspect with systems that use the very same frequency band. Other stochastic frequency allocation may ease the thrive for results, but normally coincide with lower allowance for power according to the set conditions of unlicensed usage.

Allowable time differences mostly vary with motion of the observed target. As for locating with absolute coordinates in a noisy environment several measurements are required and for disambiguation in space generally four measurements from independent reference points determine a target location, time appears as the sparse parameter.

Sequencing as a Restriction

In general, the strict sequencing of tasks appears with single tasking in one processor. Similarly the factual sequencing on only one frequency results from anti collision procedures. Both types of sequencing produce some dilation of time (with anti collision) or some dilution of location (with moving targets), while the respective wireless processes are performed by each target.

Bandwidth as a Restriction

The measuring of signals with steady modulation is bound to the bandwidth of the modulated carrier. The measuring of chirped signals is equivalently bound to the bandwidth of the transmitted pulses. In both methods, the available bandwidth will limit the precision of measurements.

Resolution as a Restriction

Technical means for measuring offer limited resolution and respective digitizing errors. This limits again the quality of results. Any way to overcome such limits raises system cost. Hence the escape again is not in improving the technical effort, but directs to the mathematical yield.

Battery Life Cycle as A Limiting Factor

The use of primary or secondary cells in wireless nodes limits both the time of operation as well as the life cycle without change for fresh batteries or just recharging. The mode of operation will be designed accordingly to widen the span of battery supply. That may be achieved by sleep up mode with respective wake up circuitry, operating without receiver in connectionless beacon mode, low repetition cycles and optimally low transmission power. An integrated loading circuit raises cost but saves the cost for external contacts. An unchangeable primary battery improves by lower self discharge compared to secondary cells, but causes the need for complete replacement or at least of the casing facultatively.

Motion as a Dynamic Challenge

When a target moves at a certain speed, the sequential measuring of distances from such transmitter target to a set of responder targets may deliver distance data for the subsequent locations at each measuring directly back to the transmitter target. This effect is independent from architecture of the network.

However, a measuring triggered from the transmitter target but performed almost in parallel by a set of receiver targets delivers a much better result under motion conditions, but requires either a server function for collecting the resulting data or requires additional response back to the triggering transceiver target.

The other escape is to apply a procedure to bundle the required measurements for each target in direct sequence thus reducing one effect of motion challenge by saving the preparation times for a reporting communications link. If not, then the competition for non-colliding transmission will lengthen the time span for each set of transmissions.

Population as Restriction

When several targets move independently in the same area or space and same wireless reach and also request locating independently and potentially in timely conflict using on the very same frequency for communications and for measurement, then the required measurements in one single ambience may collide. One escape again is the separation of the measuring and the communications processes with allocation to different frequencies, however requiring respective dual transponder capabilities.

Line of Sight as a Problem

In any case, line of sight is required for correct distance measurements. This may be eased by using auxiliary targets, but then increases the count of measurements. And the usage of auxiliary targets burdens the results with an increase of numeric inaccuracy.

Multipath Propagation as a Problem

Multipath propagation is inevitable with wireless systems. The reception from any transmitter and the response to any transmission are both challenged by the option of multiple propagation paths. If there is none but a single cranked path, there is no desired result at all, and the option to discriminate false measures from proper measures fails completely.

Typical issues with multipath propagation are fading, dither, diffraction, combining as non-linearity effects for the distance model. Additionally with power level measure the transmission through walls delivers rough errors, even with travel time measurement such error occurs.

Mathematical Requirements and Options

Mathematics serves for everything that cannot be covered by physics approaches. The assumption that a most qualified electro technical approach solves all problems arising from measurement is naïve and does not lead to sufficient results. At least thriving for best performance only at the expense of electronics is not an economized approach.

An operationally sufficient locating system will balance benefit and effort. Measurement and estimate shall take motion into account. This must not include the measuring of motion itself, but proper assessment of current and past motion to estimates. All estimate approximating the real location of the target is determined on the basis of a statistical model for the observed stochastic processes. Such model and estimation will use the set of observed propagation parameters. Some calibration data may serve as a basis for a statistical model of the propagation parameters. Such calibration is performed versus a spatial distribution of radio energy and with aspect to a known spatial distribution of corresponding targets. Other passive objects affecting propagation interfere with wireless operation and measurement.

Filtering as a Basic Requirement

Any measurement is always biased with disturbances from ambient radiation out of electrical units with switches, from other wireless units and from stationary equipment as computers. To eliminate erratic results, some estimation based on past behavior, current dynamic properties and with reference to coupling mechanisms is recommended. State of the art for such tracking is e.g. extended Kalman filtering. Approaches that do not apply filtering produce no reasonable results. However, scalar filtering uses the model of residence in a fixed location or stationary motion. If abrupt turning the direction of motion, the filter algorithm may totally fail until filtering has recovered from tracking a sufficient walk in the new direction has been performed.

Statistics as a Means for Estimation

In case of biased signals the prerequisite for filtering is some statistic estimation, which serves for eliminating the large errors and smoothes a sequence of measurement results. This may be integrated with filtering, as far as the eliminating of large errors does not bias the filtering process under any conditions.

Quadratic Equations as a Problem

The determining equations are quadratic ones, thus requiring at least one more equation (n+1) than defined by dimension (n). This leads to a minimum requirement of three equations for planar problems and four equations for spatial problems.

Over Determination as a Support

The common approach to locating calculations may be the inversion of the Euclidic distance equations. However, such deterministic approach does not serve for the balancing in over determined equation systems. The easy approach is the exploitation of Gauss' least squares principle with the multi dimensional scaling according to Torgerson.

Wireless Coexistence

Many offered systems architectures and product offerings use license free ISM frequency bands and reside in similar channel patterns. The operating of fuzzy locating shall not compromise the communications options. Some restrictions apply not to infringe this requirement.

Technology Approaches and Options

The second step after scalar calculation is the involvement of model data according to the dimensionality of the motion. If reference is made to targets in other planes but the plane on which the moving targets may operate, such model must be a three-dimensional model. For model based operations, there are several options.

Coincident Locating as the Initial Option

Imagine a worker operating with a handheld reader of any type. The person is skilled to capture the identity of an object and used reader will report the capturing with time stamp. Such report discloses the location of capture as far as this information is reported in contents. The mandatory condition could be some automatic means to capture the location at the moment of identifying. In all other cases the quality and reliability of the location report limits the validity of the e.g. vocally reported data.

In all implementations of automatic data acquisition and locating systems the option of locating a handheld reader in the moment of manual triggering shall be foreseen as the fall back option. Otherwise the robustness of automatic data acquisition systems operation is bound to availability of automatic operation only.

Choke Point Locating as the Poorest Option

A choke point is a static bottleneck in process flow designs. There the passage of individuals and/or objects may indicate the identity of such entity to a steadily installed identifier unit. This approach under all conditions is restricted to just one location.

Politics and Sales force may describe that as locating, but it is definitively still just identifying.

Power Mapping as a First Poor Option

Propagation of radio signals happens according to Maxwell's equations and includes attenuation in atmosphere proportional to distance, Such concept is the basis for power mapping. The irregularities from local ambient conditions may be taken into account by power measurement in the operational area to correct the theoretically linear attenuation with distance. However, this approach does not work with an accuracy of better than 10% of the calculated distances in the range of propagation, thus leading to accuracies in the range of some meters.

Time Distance Equivalence for Radiation

Propagation of radio signals happens according to Maxwell's equations and includes travel time in atmosphere proportional to distance, Such concept is the basis for precise distance measurement. The irregularities from local ambient conditions are not dominant, thus the approach is more precise than power measurement. So this approach works with an accuracy of better than 1% of the calculated distances in the range of propagation, thus leading to accuracies in the range of some centimeters. However, this approach serves as well for line of sight propagation as for indirect reflected propagation.

Space Model as a Strong Option

To escape the biasing with secondary paths, there must be some reasoning that excludes the physically impossible locations from sets of results from locating. Simply, all calculated locations in material will be assessed as erratic, all calculated locations at distances not possible with inherent speed limits will be assessed as erratic and all locations above ground will be assessed impossible for floor operation. The requirement for space modeling leads to depicting the operational planes different from the limits to such planes, as walls, racks, and other installations.

Statistical Model to Exploit the Measurements

There is no chance to base stable results for location on single measurements. Statistics allow for

- combining subsequent measurement results to form a track

- smoothing a single location from subsequent measurement results

- iterating stable results from coarse first estimates

The methods for computing a set of results are described in context of various applications not just with locating technologies.

Fuzzy Reasoning with Discrete Spatial Compartments

As far as locating just has to support discrimination of rooms where a target may reside, the continuous model approaches may be combined with reasoning procedures to eliminate improbable results and to exclude operationally invalid locations from potential depicting a scenery. The known methods of inference apply to such processing.

Geometric Mapping Contributions to Reasoning

As far as the ambient operational conditions are stable, a geometric mapping of the neighborhood may support the reasoning. Then all massive obstacles describe the residual space of operation. As well such mapping will support the systematic and well determined consideration of multipath propagation effects. Hence geometric mapping derives the major gain compared to Bayes' estimators.

Adaptive Approaches

A common approach as preparatory mapping requires steady conditions and a constant ambience. This crucial condition is not fulfilled in dynamic operational theatres. However, a robust solution will always detect and investigate the actual conditions and reconnoiter the present ambience. For robot navigation, the methods of adaptive systems design, hence application of learning functions, is state of the art. However, adaptation requires time. A fully adaptive solution not applying a priori knowledge will be rather slow and will show limited dynamics. A balanced combination of adaptive functions will allow for best performance in a generally known ambience and cope for all changes that occurred after last encounter.

Operational Requirements

Locating arises from operational challenges. Traditional understanding of well kept enterprises with well educated staff is undermined with a thriving for reduced skills to achieve lesser cost. In result and in addition with continuing socio-economical disparities the processes and objects under control are threatened by negligence, fraud and theft.

Evidencet

A simple indication of presence is given with the signal used for locating an object or a person. However, as presence may be temporary, a time stamp is required to adduce evidence in retrospective.

Cooperation Requirement

Persons carrying transponders or tagged objects with transponders might not be willing to be observed though having agreed earlier to this process. Then cooperation may be technically required, but individually denied. The robustness of the detection hence shall not be dependent to such cooperation. Especially covering the transponder or tearing off the transponder or otherwise tampering must be sensed automatically.

Proof of Presence

The presence of an object or a person in an operational vicinity is a strong demand. Absence of required resources generally affects planned processes. Therefore the proof of presence may be performed as far as possible before binding of additional resources happens.

Co-locating of Staff

Specially team work is bound to availability of required staff. The persons involved in a scheduled operation are well skilled to determine who is missing, but locating the missing parties is not that easy and may be strongly improved by system support.

Discrimination of Rooms

To allow for operation the respective room shall coincide with the scheduled action. Any request to operate under restrictions outside the planned confined area is suspect and may challenge security of processes and of secured knowledge. Locating the acting entities in the named confinement contributes to fulfilling security requirements.

Coincidence of Presence and Challenge

A person may try to access secured data, material or other resources outside the well secured rooms or areas. However, control may not always secure the subordination of the user to given orders. Locating simply in close distance or just in contiguity to allowed work positions may confirm the request for access as a basic feature.

Other application is coincidence of service provider, e.g.physician, with service requestor, e.g. patient, in a hospital. After identifying both persons as estimating their radial distance then access to the patient's file may be granted without any error. Such function would not be viable with precise or crisp systems as precision and allowable cost are in contradiction with absolute coordinate estimates.

Quality Requirements

The above listed terms will show that the definition of desired precision and accuracy, of repeatability and delays alone does not comply with a proper definition of requirements under aspects of cost. However, other terms of quality apply without restrictions.

Tamper Proof Identity

Basic requirement for any means to support locating is a tamper proof inherent identity of the carrying target with secured access. The secrecy of the identity prevents from plump copying threat and the tamper protection prevents from manipulating the target.

Self Identifying Authenticity

Persons who pursue to access data and applications normally authenticate themselves. Such authentication is generally bound to known locations, where the persons are authorized to perform work. Locating persons when they challenge authorization procedures is and advantage to prevent from fraud and theft.

Object Identifying Security

Numerous means are known to identify objects. Normally the location where hand held units are operated are just roughly determined by the access point where connection to network is made. However such locating is still an improvement in many operations to secure knowledge about whereabouts of objects upon identifying.

Tracking Capability

In larger context of spatially distributed services and especially in logistics, numerous objects are in use in parallel, in different locations or on the move. Especially with transportation whereabouts of objects are understood as an essential to achieve high quality of service. As far as trust is with the forwarder, no problem exists on the journey and locating may happen just on leave and upon arrival. But third party infringements may collide with this assumption and generate a demand for permanent tracking on the journey as well. Then fuzzy locating is an economized and sufficient approach, which shall not provide location data with high metric accuracies, but status information with checkable and justifiable evidence.

Tracing Capability

In case single objects are lost, the capability to trace the whereabouts is another option to get access to the missing target again. However, this tracing is performed on yet available data and no means will deliver the data from the past without respective precautions. Especially in transportation whereabouts of lost objects are understood as an essential to retrieve the missing belongings.

Alert on Deviation

Easily any deviation from planned course, set route and scheduled arrival may lead to an alert. This requires timely locating and comparison of captured data with planning.

References

- Dutton, Benjamin (2004). "15 – Basic Radio Navigation". Dutton's Nautical Navigation (15 ed.).

Naval Institute Press. pp. 154–163. ISBN 155750248X.

- Smith, D.J. (2005). Air Band Radio Handbook (8th Edition). Sutton Publishing. pp. 104–105. ISBN 0-7509-3783-1.

- Titterington, B.; Williams, D.; Dean, D. (2007). Radio Orienteering - The ARDF Handbook. Radio Society of Great Britain. ISBN 978-1-905086-27-6.

- deRosa, L.A. (1979). "Direction Finding". In J.A. Biyd; D.B. Harris; D.D. King; H.W. Welch, Jr. Electronic Countermeasures. Los Altos, CA: Peninsula Publishing. ISBN 0-932146-00-7.

- Bauer, Arthur O. (Dec 26, 2004). "Some historical and technical aspects of radio navigation, in Germany, over the period 1907 to 1945" (PDF). Retrieved 25 July 2013.

Tracking Systems: An Overview

Tracking systems offer the capability of providing real-time tracking and location management for various devices. The technology uses radio frequencies for computation and communication. Tools and techniques are an important component of any field of study. The following chapter elucidates the various tools and techniques that are related to radio frequency identification.

Tracking System

Generally a tracking system is used for the observing of persons or objects on the move and supplying a timely ordered sequence of respective location data to a model e.g. capable to serve for depicting the motion on a display capability.

An M998 High-Mobility Multipurpose Wheeled Vehicle (HMMWV) carrying a radar and tracking system shelter sits parked at an airfield during Operation Desert Shield. The shelter is used by the Marines of the 3rd Remotely Piloted Vehicle (RPV) Platoon to track their Pioneer RPVs during flight.

Tracking in Virtual Space

In virtual space technology, a tracking system is generally a system capable of rendering virtual space to a human observer while tracking the observer's body coordinates. For instance, in dynamic virtual auditory space simulations, a real-time head tracker provides feedback to the central processor, allowing for selection of appropriate head-related transfer functions at the estimated current position of the observer relative to the environment.

Tracking in Real World

There are myriad tracking systems. Some are 'lag time' indicators, that is, the data is collected after an item has passed a point for example a bar code or choke point or gate. Others are 'real-time' or 'near real-time' like Global Positioning Systems depending on how often the data is refreshed. There are bar-code systems which require a person to scan items and automatic identification (RFID auto-id). For the most part, the tracking worlds are composed of discrete hardware and software systems for different applications. That is, bar-code systems are separate from Electronic Product Code (EPC) systems, GPS systems are separate from active real time locating systems or RTLS for example, a passive RFID system would be used in a warehouse to scan the boxes as they are loaded on a truck - then the truck itself is tracked on a different system using GPS with its own features and software. The major technology "silos" in the supply chain are:

Distribution/Warehousing/Manufacturing

Indoors assets are tracked repetitively reading e.g. a barcode, any passive and active RFID and feeding read data into Work in Progress models (WIP) or Warehouse Management Systems (WMS) or ERP software. The readers required per choke point are meshed auto-ID or hand-held ID applications.

However tracking could also be capable to provide monitoring data without binding to fixed location by using a cooperative tracking capability, e.g. an RTLS.

Yard Management

Outdoors mobile assets of high value are tracked by choke point, 802.11, Received Signal Strength Indication (RSSI), Time Delay on Arrival (TDOA), active RFID or GPS Yard Management; feeding into either third party yard management software from the provider or to an existing system. Yard Management Systems (YMS) couple location data collected by RFID and GPS systems to help Supply Chain Managers to optimize utilization of yard assets such as trailers and dock doors. YMS systems can use either active or passive RFID tags.

Fleet Management

Fleet management is applied as a tracking application using GPS and composing tracks from subsequent vehicle's positions. Each vehicle to be tracked is equipped with a GPS receiver and relays the obtained coordinates via cellular or satellite networks to a home station. Fleet management is required by:

- Large fleet operators, (vehicle/railcars/trucking/shipping)

- Forwarding operators (containers, machines, heavy cargo, valuable shippings)

- Operators who have high equipment and/or cargo/product costs

- Operators who have a dynamic workload

Attendance Management

One such use of the RFID technology is in tracking IDs of students. Using GPS IDs would resolve the decreasing attendance in schools by monitoring the whereabouts of students when they did not attend class (Jensen, 2008). It is also used to efficiently check attendance. Perks of this tracking system is allowing students to check out library books buy food in the cafeterias (Jensen, 2008). The GPS IDs also act as a security measure to monitor any unwanted visitors or an emergency locator if a student cannot be found (Jensen, 2008). In the Spring Independent School District, students have been using for many years in check that students are staying in school during the day. Since they have instigated the system, attendance has increased thus schooling funding has increased as well (Jensen, 2008).

Recently, debates over the Fourth Amendment have come up. Conservative students wish to keep their privacy and forbid to wear tracking devices, especially hackers can break into these systems to find out students' information. Since many schools, such as those in the Spring Independent School District, require students to wear the tracking IDs, students argue that it is an immediate violation of their privacy (Jensen, 2008). Yet, the Fourth Amendment is not violated in these cases since students are not tracked in their homes (Warner, 2007). Each school's decision over GPS IDs varies as states develop laws against these IDs in schools and as students protest for their privacy rights.

Mobile Phone Services

Location-based services or LBS is a term that is derived from the telematics and telecom world. The combination of A-GPS, newer GPS and cellular locating technology is what has enabled the latest "LBS" for handsets and PDAs. Line of sight is not necessarily required for a location fix. This is a significant advantage in certain applications since a GPS signal can still be lost indoors. As such, A-GPS enabled cell phones and PDAs can be located indoors and the handset may be tracked more precisely. This enables non-vehicle centric applications and can bridge the indoor location gap, typically the domain of RFID and RTLS systems, with an off the shelf cellular device.

Currently, A-GPS enabled handsets are still highly dependent on the Location-Based Service (LBS) carrier system, so handset device choice and application requirements are still not apparent. Enterprise system integrators need the skills and knowledge to correctly choose the pieces that will fit the application and geography.

Operational Requirements

Regardless of the tracking technology, for the most part the end-users just want to locate themselves or wish to find points of interest. The reality is that there is no "one size fits all" solution with locating technology for all conditions and applications.

Application of tracking is a substantial basis for vehicle tracking in fleet management, asset management, individual navigation, social networking, or mobile resource management and more. Company, group or individual interests can benefit from more than one of the offered technologies depending on the context.

GPS Applications

GPS has global coverage but can be hindered by line-of-sight issues caused by buildings and urban canyons. RFID is excellent and reliable indoors or in situations where close proximity to tag readers is feasible, but has limited range and still requires costly readers. RFID stands for Radio Frequency Identification. This technology uses electromagnetic waves to receive the signal from the targeting object to then save the location on a reader that can be looked at through specialized software (Warner, 2007).

Real-time Locating Systems (RTLS)

RTLS are enabled by Wireless LAN systems (according to IEEE 802.11) or other wireless systems (according to IEEE 802.15) with multilateration. Such equipment is suitable for certain confined areas, such as campuses and office buildings. RTLS require system-level deployments and server functions to be effective. RTLS systems are affordable and accurate for industrial and yard applications. RTLS systems are not appropriate for all indoor applications, there fuzzy locating systems with unilateration may apply more economically.

Chipless RFID

Chipless RFID tags are RFID tags that do not require a microchip in the transponder.

RFIDs offer longer range and ability to be automated, unlike barcodes that require a human operator for interrogation. The main challenge to their adoption is the cost of RFIDs. The design and fabrication of ASICs needed for RFID are the major component of their cost, so removing ICs altogether can significantly reduce its cost. The major challenges in designing chipless RFID is data encoding and transmission.

Communication Techniques

Time-domain Reflectometry Vs Frequency Signature Devices

Chipless RFID tags may use either time-domain reflectometry or frequency signature techniques. In time domain reflectometry the interrogator sends a pulse and listens for echoes. The timing of pulse arrivals encodes the data. In frequency signature RFIDs the interrogator sends waves of several frequencies, a broadband pulse or a chirp, and monitors the echoes' frequency content. The presence or absence of certain frequency components in the received waves encodes the data. They may use chemicals, magnetic materials or resonant circuits to attenuate or absorb radiation of a particular frequency.

Chemical-based

Self-generating Ceramic Mixtures

In 2001, Roke Manor Research centre announced materials that emit characteristic radiation when moved. These may be exploited for storage of a few data bits encoded in the presence or absence of certain chemicals.

Biocompatible Ink

Somark employed a dielectric barcode that may be read using microwaves. The dielectric material reflects, transmits and scatters the incident radiation; the different position and orientation of these bars affects the incident radiation differently and thus encodes the spatial arrangement in the reflected wave. The dielectric material may be dispersed in a fluid to create a dielectric ink. They were mainly used as tags for cattle, which were "painted" using a special needle. The ink may be visible or invisible according to the nature of the dielectric, Operating frequency of the tag may be changed by using different dielectrics.

CrossID Nanometric Ink

This system uses varying magnetism. Materials resonate at different frequencies when excited by radiation. The reader analyzes the spectrum of the reflected signal to identify the materials. 70 different materials were found. Each material's presence or absence may be used to encode a bit, enabling encoding up to 2^{70} unique binary strings. They work on frequencies between three and ten gigahertz.

Passive Antenna

In 2004 Tapemark announced a chipless RFID that will have only a passive antenna with a diameter as small as 5 μm. The antenna consists of small fibers called nano-resonant structures. Spatial difference in structure encode data. The interrogator sends out a coherent pulse and reads back an interference pattern that it decodes to identify a tag. They work from 24 GHz–60 GHz. Tapemark later discontinued this technology.

Magnetism-based

Programmable Magnetic Resonance

Sagentia's devices are acousto-magnetic. They exploit the resonance features of magnetically soft magnetostrictive materials and the data retention capability of hard magnetic materials. Data is written to the card using the contact method. The resonance of the magnetostrictive material is altered by the data stored in the hard material. Harmonics may be enabled or disabled corresponding to the state of the hard material, thus encoding the device state as a spectral signature. Tags built by Sagentia for Astra-Zeneca fall into this category.

Magnetic Data Tagging

Flying Null technology uses a series of passive magnetic structures, much like the lines used in conventional barcodes. These structures are made of soft magnetic material. The interrogator contains two permanent magnets with like poles. The resulting magnetic field has a null volume in the centre. Additionally, interrogating radiation is used. The magnetic field created by the interrogator is such that it drives the soft material to saturation except when it is at the null volume. When in the null volume the soft magnet interacts with the interrogating radiation thus giving away the position of the soft material. Spatial resolution of more than 50 μm may be attained.

Surface Acoustic Wave

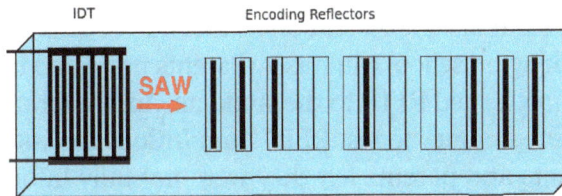

Illustration of a simple SAW RFID encoding 013 in base 4. The first and last reflectors are used for calibration. The second and second last for error detection. The data is encoded in the remaining three groups. Each group contains 4 slots and an empty slot followed by another group.

Surface acoustic wave devices consists of a piezoelectric crystal-like lithium niobate on which transducers are made by single-metal-layer photolithographic technology. The transducers usually are Inter-Digital Transducers (IDT), which have a two-toothed comb-like structure. An antenna is attached to the IDT for reception and transmission. The transducers convert the incident radio wave to surface acoustic waves that travel on the crystal surface until it reaches the encoding reflectors that reflect some waves and transmit the rest. The IDT collects the reflected waves and transmits them to the reader. The first and last reflectors are used for calibration as the response may be affected by physical parameters such as temperature. A pair of reflectors may also be used for error correction. The reflections increase in size from nearest to farthest of the IDT to account for losses due to preceding reflectors and wave attenuation. Data is encoded using Pulse Position Modulation (PPM). The crystal is logically divided into groups, such that each group typically has a length equal to the inverse of the bandwidth. Each group is divided into slots of equal width. The reflector may be placed in any slot. The last slot in each group is usually unused, leaving n-1 positions for the reflector, thus encoding n-1 states. The repetition rate of the PPM is equal to the system bandwidth. The reflector's slot position may be used to encode phase. The devices' temperature dependence means they can also act as temperature sensors.

Capacitively Tuned Split Microstrip Resonators

They employ a grid of dipole antennas that are tuned to different frequencies. The interrogator generates a frequency sweep signal and scans for signal dips. Each dipole antenna can encode one bit. The frequency swept will be determined by the antenna length.

RFID in Schools

Various schools have been using radio-frequency identification technology to record and monitor students.

United States

It is thought that the first school in the USA to introduce RFID technology was Spring Independent School District near Houston, Texas. In 2004, it gave 28,000 students RFID badges to record when students got on and off school buses. This was expanded in 2008 to include location tracking on school campuses. Parents protested in January 2005 when Brittan Elementary School issued RFID to the students. Administrators at a school in Sutter, California, were offered money to test RFID from InCom and issued RFID-chipped ID tags to students. Students and parents felt they were not fully informed about the RFID and questioned the tactics the school used to implement the program, and the ethics of the monetary deal the school made with the company to test and promote its product. Parents quickly squashed the program with help from the American Civil Liberties Union.

In 2012, Northside Independent School District, San Antonio, Texas introduced active RFID, worn on a lanyard around the student's necks. One student refused to participate in the program and was expelled from school, after a court case. The school eventually dropped the RFID program and started tracking students with cameras instead.

UK

In 2007, Hungerhill High School, Doncaster, UK, tried RFID chips sewn into students' blazers. Ten children tested the RFID for attendance. There were privacy concerns, and the trial was stopped.

West Cheshire College integrated active ultra wideband (UWB) RFID into their new college campuses in Chester in 2010, and Ellesmere Port in 2011, to tag students and assets using a real time location system (RTLS). Students wore the active RFID tags around their necks. West Cheshire College stopped RFID tagging students in February 2013. A series of Freedom of Information requests were sent to the college about the RFID tracking of students. Specifications of the active RFID at West Cheshire College:

- Ultra wideband RFID tags emit brief radio frequency signals across the entire 6.35 to 6.75 GHz frequency band.

- Average battery lifespan of a RFID tag is seven years.

- Receivers, which can receive tag signals up to 328 feet away, are located throughout the campus buildings, in order to ensure that the tags can be pinpointed regardless of where within the school a student might be.

- RFID tags provides accuracy to within 1 meter (3.3 feet).

- RFID Tag transmission rate of once per second.

- West Cheshire College uses RFID with a real time location system.

- The real time location system enables observation of student and staff in peer groups.

Germany

After a school shooting in Germany in 2009, which claimed 16 lives, the Friedrich-von-Canitz School implemented a real-time location technology over Wi-Fi. The solution was developed by the German company How To Organize (H2O) GmbH in cooperation with teachers and the local police force.

Passive RFID

Passive RFID is used routinely in schools to register teachers and students and to provide access to services such as photocopying and door access.

GPS Tracking Unit

A GPS tracking unit is a device, normally carried by a moving vehicle or person, that uses the Global Positioning System to determine and track its precise location, and hence that of its carrier, at intervals. The recorded location data can be stored within the tracking unit, or it may be transmitted to a central location data base, or Internet-connected computer, using a cellular (GPRS or SMS), radio, or satellite modem embedded in the unit. This allows the asset's location to be displayed against a map backdrop either in real time or when analysing the track later, using GPS tracking software. Data tracking software is available for smartphones with GPS capability.

GPS Tracking Unit Architecture

A GPS tracker essentially contains a GPS module to receive the GPS signal and calculate the coordinates. For data loggers it contains large memory to store the coordinates, data pushers additionally contains the GSM/GPRS modem to transmit this information to a central computer either via SMS or via GPRS in form of IP packets.

Types of GPS Trackers

Usually, a GPS tracker will fall into one of these three categories, though most GPS-equipped phones can work in any of these modes according to mobile applications installed:

Data Loggers

A GPS logger simply logs the position of the device at regular intervals in its internal memory. Modern GPS loggers have either a memory card slot, or internal flash memory card and a USB port. Some act as a USB flash drive. This allows downloading of the track log data for further analysis in a computer. The tracklist or point of interest list may be in GPX, KML, NMEA or other format.

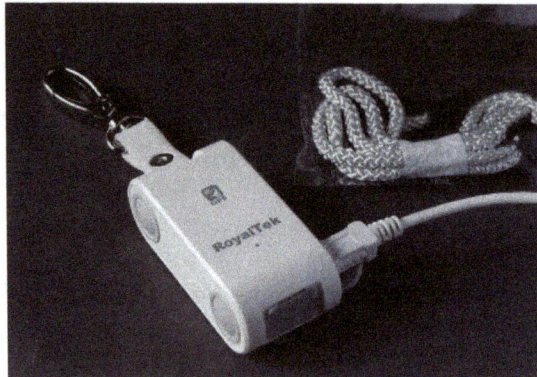

Typical GPS logger

Most digital cameras save the time a photo was taken. Provided the camera clock was reasonably accurate or used GPS as its time source, this time can be correlated with GPS log data, to provide an accurate location. This can be added to the Exif metadata in the picture file. Cameras with GPS receiver built in can directly produce such a geo-tagged photograph.

In some private investigation cases, data loggers are used to keep track of a target vehicle. The PI need not follow the target so closely, and always has a backup source of data.

Data Pushers

Data pusher is the most common type of GPS tracking unit, used for asset tracking, personal tracking and Vehicle tracking system.

Also known as a GPS beacon, this kind of device pushes (i.e. "sends") the position of the device as well as other information like speed or altitude at regular intervals, to a determined server, that can store and instantly analyze the data.

A GPS navigation device and a mobile phone sit side-by-side in the same box, powered by the same battery. At regular intervals, the phone sends a text message via SMS or GPRS, containing the data from the GPS receiver. Newer GPS-integrated smartphones running GPS tracking software can turn the phone into a data pusher (or logger) device; as of 2009 open source and proprietary applications are available for common Java ME enabled phones, iPhone, Android, Windows Mobile, and Symbian.

Most 21st-century GPS trackers provide data "push" technology, enabling sophisticated GPS tracking in business environments, specifically organizations that employ a mobile workforce, such as a commercial fleet. Typical GPS tracking systems used in commercial fleet management have two core parts: location hardware (or tracking device) and tracking software. This combination is often referred to as an Automatic Vehicle Location system. The tracking device is most often hardwire installed in the vehicle; connected to the CAN-bus, Ignition system switch, battery. It allows collection of extra data, which later gets transferred to the GPS tracking server, where it is available for viewing, in most cases via a website accessed over the internet, where fleet activity can be viewed live or historically using digital maps and reports.

GPS tracking systems used in commercial fleets are often configured to transmit location and telemetry input data at a set update rate or when an event (door open/close, auxiliary equipment on/off, geofence border cross) triggers the unit to transmit data. Live GPS Tracking used in commercial fleets, generally refers to systems which update regularly at 1 minute, 2 minute or 5 minute intervals, whilst the ignition status is on. Some tracking systems combine timed updates with heading change triggered updates.

GPS tracking solutions are recently being used in mainstream commercial auto insurance these are sometimes called Telematics 2.0.

The applications of trackers of this kind include:

Personal Tracking

- Law enforcement. An arrested suspect out on bail may have to wear a GPS tracker, usually an ankle monitor, as a bail condition. GPS tracking may also be ordered for persons subject to a restraining order.

- Race control. In some sports, such as gliding, participants are required to carry a tracker. In particular this allows race officials to know if the participants are cheating, taking unexpected shortcuts, and how far apart they are. This use was illustrated in the movie Rat Race.

- Espionage/surveillance. A tracker on a person or vehicle allows movements to be tracked. This application is used by private investigators.

- These devices are also used by some parents to track their children. Some devices can send text alerts to the parents for every unexpected place visited by the child.

- GPS personal tracking devices are used in the care of the elderly and vulnerable. Devices allow users to call for assistance and optionally allow designated carers to locate the user's position, typically within 5 to 10 metres. Their use helps promote independent living and social inclusion for the elderly. Devices often incorporate either 1-way or 2-way voice communication. Some devices also allow the user to call several phone numbers using pre-programmed speed dial buttons. Trials using GPS personal tracking devices are also underway in several countries for use with early stage dementia.

- Some Internet Web 2.0 pioneers have created their own personal web pages that show their position constantly, and in real time, on a map within their website. These usually use data push from a GPS enabled cell phone or a personal GPS tracker.

- Sports: the movements of a rambler, cyclist, etc., can be tracked. Statistics such as instantaneous and average speed, and distance traveled, are logged.

- Adventure sports: GPS tracking devices such as the SPOT Satellite Messenger are available that allow the position of a person to be tracked. In particular this allows rescue personnel to locate the carrier. These devices allow the carrier to send messages, even when out of cellular telephone range.

- Monitoring employees. GPS-handled tracking devices with built-in cellphone are used to monitor employees by various companies, specially those engaged in field work.

Asset Tracking

- Solar Powered. The advantage of some solar powered units is that they have much more power over their lifetime than battery powered units. This gives them the advantage to report their position and status much more often than battery units which need to conserve their energy to extend their life. Some wireless solar powered units, such as the RailRider can report more than 20,000 times per year and work indefinitely on solar power eliminating the need to change batteries.

- Animal control. When put on a wildlife animal (e.g. in a collar), it allows scientists to study its activities and migration patterns. Vaginal implant transmitters mark the location where pregnant females give birth. Animal tracking collars may also be put on domestic animals, to locate them in case they get lost.

Aircraft Trackers

Aircraft can be tracked either by ADS-B (primarily powered aircraft or by FLARM data packets picked up by a network of ground stations both of which are data pushers. ADS-B is to be superseded by ADS-C, a data puller.

Data Pullers

GPS data pullers are also known as GPS transponders. Unlike data pushers that send the position of the devices at regular intervals (push technology), these devices are always on, and can be queried as often as required (pull technology). This technology is not in widespread use, but an example of this kind of device is a computer connected to the Internet and running gpsd.

These can often be used in the case where the location of the tracker will only need to be known occasionally e.g. placed in property that may be stolen, or that does not have constant source of energy to send data on a regular basis, like freights or containers.

Data Pullers are coming into more common usage in the form of devices containing a GPS receiver and a cell phone which, when sent a special SMS message reply to the message with their location.

Covert GPS Trackers

Covert GPS trackers contain the same electronics as regular GPS trackers but are constructed in such a way as to appear to be an everyday object. One use for covert GPS trackers is for power tool protection, these devices can be concealed within power tool boxes and traced if theft occurs.

United States Law

In the US, the use of GPS trackers by government authorities is limited by the 4th Amendment to the United States Constitution, so police, for example, usually require a search warrant in most circumstances. While police have placed GPS trackers in vehicles without warrant, this usage was questioned in court in early 2009.

Use by private citizens is regulated in some states, such as California, where California Penal Code Section 637.7 states: (a) No person or entity in this state shall use an electronic tracking device to determine the location or movement of a person. (b) This section shall not apply when the registered owner, lesser, or lessee of a vehicle has consented to the use of the electronic tracking device with respect to that vehicle. (c) This section shall not apply to the lawful use of an electronic tracking device by a law enforcement agency. (d) As used in this section, "electronic tracking device" means any device attached to a vehicle or other movable thing that reveals its location or movement by transmission of electronic signals. (e) A violation of this section is a misdemeanor. (f) A violation of this section by a person, business, firm, company, association, partnership, or corporation licensed under Division 3 (commencing with Section 5000) of the Business and Professions Code shall constitute grounds for revocation of the license issued to that person, business, firm, company, association, partnership, or corporation, pursuant to the provisions that provide for the revocation of the license as set forth in Division 3 (commencing with Section 5000) of the Business and Professions Code.

Note that 637.7 pertains to all electronic tracking devices, and does not differentiate between those that rely on GPS technology or not. As the laws catch up with the times, it is plausible that all 50 states will eventually enact laws similar to those of California.

Other laws, like the common law invasion of privacy tort as well as state criminal wiretapping statutes (for example, the wiretapping statute of the Commonwealth of Massachusetts, which is extremely restrictive) potentially cover the use of GPS tracking devices by private citizens without consent of the individual being so tracked. Privacy can also be a problem when people use the devices to track the activities of a loved one. GPS tracking devices have also been put on religious statues in order to track the whereabouts of the statue if stolen.

In 2009, debate ensued about a Georgia proposal to outlaw hidden GPS tracking, with an exception for law enforcement officers but not for private investigators.

United Kingdom Law

The law in the UK has not specifically addressed the use of GPS Trackers, but several laws may affect the use of this technology as a surveillance tool.

Data Protection Act 1998

It is quite clear that if client instructions (written or digitally transmitted) that identify a person and a vehicle are combined with a tracker, the information gathered by the tracker becomes personal data as defined by the Data Protection Act 1998. The document "What is personal data? – A quick reference guide" published by the Information Commissioner's Office (ICO) makes clear that data identifying a living individual is personal data. If a living individual can be identified from the data, with or without additional information that may become available, is personal data.

Identifiability

An individual is identified if distinguished from other members of a group. In most cases an individual's name together with some other information will be sufficient to identify them; but a person can be identified even if their name is not known. Start by looking at the means available to identify an individual and the extent to which such means are readily available to you.

Does The Data 'Relate to' The Identifiable Living Individual, Whether in Personal or Family Life, Business or Profession?

Relates to' means: Data which identifies an individual, even without an associated name, may be personal data which is processed to learn or record something about that individual, or the processing of information that affects the individual. Therefore, data may 'relate to' an individual in several different ways.

Is The Data 'Obviously About' a Particular Individual?

Data 'obviously about' an individual will include his medical history, criminal record, record of his work, or his achievements in a sporting activity. Data that is not 'obviously about' a particular individual may include information about his activities. Data such as personal bank statements or itemized telephone bills will be personal data about the individual operating the account or contracting for telephone services. Where data is not 'obviously about' an identifiable individual it may be helpful to consider whether the data is being processed, or could easily be processed, to learn, record or decide something about an identifiable individual. Information may be personal data where the aim, or an incidental consequence, of the processing, is that you learn or record something about an identifiable individual, or the processing could affect an identifiable individual. Data from a Tracker would be to identify the individual or his activities. It is therefore personal data within the meaning of the Data Protection Act 1998.

Any individual who wishes to gather personal data must be registered with the Information Commissioner's Office (ICO) and have a DPA number. It is a criminal offence to process data and not have a DPA number.

Trespass

It may be a civil trespass to deploy a tracker onto a car not belonging to your client or to yourself. But in the OSC's annual inspection, the OSC's Chief Surveillance Commissioner Sir Christopher Rose stated "putting an arm into a wheel arch or under the frame of a vehicle is straining the concept of trespass".

However, entering private land of anyone in order to deploy a tracker is clearly a trespass which is a civil tort.

Prevention of Harassment Act 1997

At times, the public misinterprets surveillance, in all its forms, as stalking. Whilst there is no strict legal definition of 'stalking', neither is there specific legislation to address this behaviour. Rather, it is a term used to describe a particular kind of harassment. Generally, it is used to describe a long-term pattern of persistent and repeated contact with, or attempts to contact, a particular victim.

The Protection of Freedoms Act 2012 created two new offences of stalking by inserting new sections 2A and 4A into the PHA 1997. The new offences which came into force on 25 November 2012, are not retrospective. Section 2A (3) of the PHA 1997 sets out examples of acts or omissions which, in particular circumstances, are ones associated with stalking. Examples are: following a person, watching or spying on them, or forcing contact with the victim through any means, including social media.

Such behaviour curtails a victim's freedom, leaving them feeling that they constantly have to be careful. In many cases, the conduct might appear innocent (if considered in isolation), but when carried out repeatedly, so as to amount to a course of conduct, it may then cause significant alarm, harassment or distress to the victim.

It should be noted that the examples given in section 2A (3) is not an exhaustive list but an indication of the types of behaviour that may be displayed in a stalking offence.

Stalking and harassment of another or others can include a range of offences such as those under: the Protection from Harassment Act 1997; the Offences Against the Person Act 1861; the Sexual Offences Act 2003; and the Malicious Communications Act 1988.

Examples of the types of conduct often associated with stalking include: direct communication; physical following; indirect contact through friends, colleagues, family or technology; or, other intrusions into the victim's privacy. The behaviour curtails a victim's freedom, leaving them feeling that they constantly have to be careful.

If the subject of enquiry is aware of the tracking, then this may amount to harassment under the Prevention of Harassment Act 1997. There is a case at the Royal Courts of Justice where a private investigator is being sued under this act for the use of trackers. In December 2011, a Claim was brought against Richmond Day & Wilson Limited (First Defendant) and Bernard Matthews Limited (Second Defendant), Britain's leading Turkey Provider.

The case relates to the discovery of a tracking device found in August 2011 on a vehicle supposedly connected to Hillside Animal Sanctuary.

Regulation of Investigatory Powers Act 2000

Property Interference: The Home Office published a document entitled "Covert Surveillance and Property Interference, Revised Code of Practice, Pursuant to section 71 of the Regulation of Investigatory Powers Act 2000" where it suggests in Chapter 7, page 61 that;

General Basis for Lawful Activity

7. 1 Authorisations under section 5 of the 1994 Act or Part III of the 1997 Act should be sought wherever members of the intelligence services, the police, the services police, Serious and Organised Crime Agency (SOCA), Scottish Crime and Drug Enforcement Agency (SCDEA), HM Revenue and Customs (HMRC) or Office of Fair Trading (OFT), or persons acting on their behalf, conduct entry on, or interference with, property or with wireless telegraphy that would be otherwise unlawful.

7. 2 For the purposes of this chapter, "property interference" shall be taken to include entry on, or interference with, property or with wireless telegraphy.

Example: The use of a surveillance device for providing information about the location of a vehicle may involve some physical interference with that vehicle as well as subsequent directed surveillance activity. Such an operation could be authorised by a combined authorisation for property interference (under Part III of the 1997 Act) and, where appropriate, directed surveillance (under the 2000 Act). In this case, the necessity and proportionality of the property interference element of the authorisation would need to be considered by the appropriate authorising officer separately to the necessity and proportionality of obtaining private information by means of the directed surveillance.

This can be interpreted to mean that placing a tracker on a vehicle without the consent of the owner is illegal unless you obtain authorisation from the Surveillance Commissionaire under the RIPA 2000 laws. Since a member of the public cannot obtain such authorisations, it is therefore illegal property interference.

Another interpretation is that it is illegal to do so IF you are acting under the instruction of a public authority and you do not obtain authorisation. The legislation makes no mention of property interference for anyone else.

The second interpretation appears to be the valid one. Currently there is no legislation in place that deals with the deployment of trackers in a criminal sense except RIPA 2000 and that RIPA 2000 ONLY applies to those agencies and persons mentioned in it.

Uses in Marketing

In August, 2010, Brazilian company Unilever ran a promotion where GPS trackers were placed in boxes of Omo laundry detergent. Teams would then track consumers who purchased the boxes of detergent to their homes where they would be awarded with a prize for their purchase. The company also launched a website (in Portuguese) to show the approximate location of the winners' homes.

Fleet Management

Fleet management includes commercial motor vehicles such as cars, aircraft (planes, helicopters etc.), ships, vans and trucks, as well as rail cars. Fleet (vehicle) management can include a range of functions, such as vehicle financing, vehicle maintenance, vehicle telematics (tracking and diagnostics), driver management, speed management, fuel management and health and safety management. Fleet Management is a function which allows companies which rely on transportation in business to remove or minimize the risks associated with vehicle investment, improving efficiency, productivity and reducing their overall transportation and staff costs, providing 100% compliance with government legislation (duty of care) and many more. These functions can be dealt with by either an in-house fleet-management department or an outsourced fleet-management provider. According to market research from the independent analyst firm Berg Insight, the number of fleet management units deployed in commercial fleets in Europe will grow from 3.05 million units at the end of 2012 to 6.40 million in 2017. Even though the overall penetration level is just a few percent, some segments such as road transport will attain adoption rates above 31 percent.

Vehicle Tracking

The most basic function in all fleet management systems, is the vehicle tracking component. This component is usually GPS-based, but sometimes it can be based on GLONASS or a cellular triangulation platform. Once vehi-cle location, direction and speed are determined from the GPS components, additional tracking capabilities transmit this information to a fleet management software applica-tion. Methods for data transmission include both terrestrial and satellite. Satellite track-ing communications, while more expensive, are critical if vehicle tracking is to work in remote environments without interruption. Users can see actual, real-time locations of their fleet on a map. This is often used to quickly respond on events in the field.

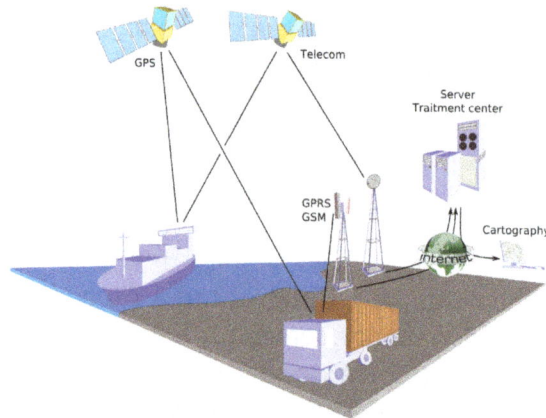

Principle of geolocation based on the GPS for the position determination and the GSM/GPRS or telecommunication satellites network for the data transmission.

Mechanical Diagnostics

Advanced fleet management systems can connect to the vehicle's onboard computer, and gather data for the user. Details such as mileage and fuel consumption are gathered into a global statistics scheme.

Driver Behavior

By combining received data from the vehicle tracking system and the on-board computer, it is possible to form a profile for any given driver (average speed, frequency of detours, breaks, etc.). All this data about vehicle behaviour might be apply for operating skills elaboration with high economical impact in mining operations. With high developed FMS mine management has a full range of data in real-time: speeds, dumping engine speed, driving with a raised body, the use of brakes, abusive shifting etc. Collected and structured data may be applied as an KPI indicator or learning outcomes tracking.

Fleet Management Software

Fleet management software enables people to accomplish a series of specific tasks in the management of any or all aspects relating to a company's fleet of vehicles. These specific tasks encompass all operations from vehicle acquisition to disposal. Software, depending on its capabilities, allows functions such as driver and vehicle profiling, trip profiling, dispatch, vehicle efficiency, etc. It can provide remote control features, such as Geo-fencing and active disabling. Current vehicle diagnostic information can also be related to a management site, depending on the type of hardware installed in the vehicles. New platform, based on Fleet management software, is fleet controlling with higher amount of information available for both drivers and dispatchers of a fleet. At this time (2012) online software platforms are very popular: users no longer have to install software and they can access the software through a web browser..

Management of Ships

Fleet management also refers to the management of ships while at sea. Shipping fleet management contracts are normally given to fleet management companies that handle aspects like crewing, maintenance, and day-to-day operations. This gives the ship owner time to concentrate on cargo booking.

Fleet Security and Control

Recent advances in fleet management allow for the addition of over-the-air (OTA) security and control of fleet vehicles. Fleet Security and Control includes security of the vehicle while stopped or not in operation and the ability to safely disable a vehicle while in operation. This allows the fleet manager to recover stolen or rogue vehicles while reducing the chance of lost or stolen cargo. The additional of Fleet Security and Control to a fleet management system gives a fleet card manager preventative measures to address cargo damage and loss.

Remote Vehicle Disabling Systems

Remote vehicle disabling systems provide users at remote locations the ability to prevent an engine from starting, prevent movement of a vehicle, and to stop or slow an operating vehicle. Remote disabling allows a dispatcher or other authorized personnel to gradually decelerate a vehicle by downshifting, limiting the throttle capability, or bleeding air from the braking system from a remote location. Some of these systems provide advance notification to the driver that the vehicle disabling is about to occur. After stopping a vehicle, some systems will lock the vehicle's brakes or will not allow the vehicle's engine to be restarted within a certain timebound.

Remote disabling systems can also be integrated into a remote panic and emergency notification system. In an emergency, a driver can send an emergency alert by pressing a panic button on the dashboard, or by using a key-fob panic button if the driver is within close proximity of the truck. Then, the carrier or other approved organization can be remotely alerted to allow a dispatcher or other authorized personnel to evaluate the situation, communicate with the driver, and/or potentially disable the vehicle.

Fleet Replacement and Lifecycle Management

The timely replacement of vehicles and equipment is a process that requires the ability to predict asset lifecycles based on costing information, utilization, and asset age. Organizations prefer to use new fleet as a strategy for cost reduction where the used fleet is sold so that a new fleet is maintained.

Funding requirements are also an issue, because many organizations, especially government, purchase vehicles with cash. The ad hoc nature and traditional low funding levels with cash has put many operations in an aged fleet. This lack of adequate funding for replacement can also result in higher maintenance costs due to aged vehicles.

Duty of Care

In the UK, in April 2008, the Corporate Manslaughter Act was strengthened to target company directors as well as their drivers in cases of road deaths involving vehicles used on business. The Police have said they now treat every road death as 'an unlawful killing' and have the power to seize company records and computers during their investigations. They will bring prosecutions against company directors who fail to provide clear policies and guidance for their employees driving at work.Unfortunately, in the UK a number of businesses are failing to meet their duty of care.In particular prosecutions can be brought against company directors for failing to meet their duty of care and allowing HGV driver hours to exceed the legal limits. Failure to comply with EU rules can result in a fixed penalty of up to £300, a graduated deposit of up to £1500 or you could be summoned to court. Directors and business owners may not be aware that privately owned vehicles used for business journeys are treated exactly the same as company owned vehicles. Directors have an equal responsibility under the law to ensure these vehicles are also roadworthy and correctly insured. It is vital that every company has a 'Driving at Work' policy in place covering every element of their business vehicle operation, no matter how few vehicles are involved and who owns them. Every employee driving for business is required to sign up to the policy. In this way the directors can reduce the risk of being prosecuted and a possible custodial sentence.

References

- "What is personal data? – A quick reference guide". ICO.ORG.UK. Information Commissioners Office. Retrieved 15 September 2014.

- "Government officials track cars and trespass on private property, report shows". The Telegraph. September 21, 2009. Retrieved 14 September 2014.

- "Animal campaigner claims car was bugged by Bernard Matthews". The Mirror Newspaper. December 22, 2011. Retrieved 14 September 2014.

- "The Use of Flying Null Technology in the Tracking of Labware in Laboratory Automation". JALA. Retrieved 17 August 2013.

- Schofield, Jack (2005-02-10). "School RFID Plan Gets an F | Technology". theguardian.com. Retrieved 2013-10-12.

- "US Pupil Andrea Hernandez Loses 'Mark Of The Beast' Tracking Badge Appeal". Huffingtonpost.co.uk. Retrieved 2013-10-12.

- "West Cheshire College Tracks Whereabouts of Students, Staff - RFID Journal". Web.archive.org. Archived from the original on 2012-09-15. Retrieved 2013-10-12.

- Clancy, Heather. "California security company uses barcodes to help track assets". CBS Interactive. Retrieved February 9, 2012.

- Hadlock, Charles (Oct 14, 2012). "RFID chips let schools track students -- and retain funding -- but some parents object". NBC News. Retrieved 22 April 2014.

- Koch, Wendy (2009-02-12). "Cheatin' hearts pump up economy on Valentine's Day". USA Today. Retrieved 2010-05-04.

Data Identification Techniques: A Comprehensive Study

Data identification techniques refer to those devices that use radio frequencies to transmit and receive information as well as compute and make decisions with that data. The topics in this section discuss devices as well standards of data identification and processing. The aspects elucidated in this chapter are of vital importance, and provide a better understanding of radio frequency identification.

Automatic Identification and Data Capture

Automatic identification and data capture (AIDC) refers to the methods of automatically identifying objects, collecting data about them, and entering that data directly into computer systems (i.e. without human involvement). Technologies typically considered as part of AIDC include bar codes, Radio Frequency Identification (RFID), biometrics, magnetic stripes, Optical Character Recognition (OCR), smart cards, and voice recognition. AIDC is also commonly referred to as "Automatic Identification," "Auto-ID," and "Automatic Data Capture."

AIDC is the process or means of obtaining external data, particularly through analysis of images, sounds or videos. To capture data, a transducer is employed which converts the actual image or a sound into a digital file. The file is then stored and at a later time it can be analyzed by a computer, or compared with other files in a database to verify identity or to provide authorization to enter a secured system. Capturing of data can be done in various ways; the best method depends on application.

AIDC also refers to the methods of recognizing objects, getting information about them and entering that data or feeding it directly into computer systems without any human involvement. Automatic identification and data capture technologies include barcodes, RFID, bokodes, OCR, magnetic stripes, smart cards and biometrics (like iris and facial recognition system).

In biometric security systems, capture is the acquisition of or the process of acquiring and identifying characteristics such as finger image, palm image, facial image, iris print or voice print which involves audio data and the rest all involves video data.

Radio-frequency identification (RFID) is relatively a new AIDC technology which was first developed in 1980s. The technology acts as a base in automated data collection, identification and analysis systems worldwide. RFID has found its importance in a wide range of markets including livestock identification and Automated Vehicle Identification (AVI) systems because of its capability to track moving objects. These automated wireless AIDC systems are effective in manufacturing environments where barcode labels could not survive.

Capturing Data from Printed Documents

One of the most useful application tasks of data capture is collecting information from paper documents and saving it into databases (CMS, ECM and other systems). There are several types of basic technologies used for data capture according to the data type:

- OCR – for printed text recognition

- ICR – for hand-printed text recognition

- OMR – for marks recognition

- OBR – for barcodes recognition

- BCR – for bar code recognition

- DLR - for document layer recognition

These basic technologies allow extracting information from paper documents for further processing it in the enterprise information systems such as ERP, CRM and others.

The documents for data capture can be divided into 3 groups: structured, semi-structured and unstructured.

Structured documents (questionnaires, tests, insurance forms, tax returns, ballots, etc.) have completely the same structure and appearance. It is the easiest type for data capture, because every data field is located at the same place for all documents.

Semi-structured documents (invoices, purchase orders, waybills, etc.) have the same structure but their appearance depends on number of items and other parameters. Capturing data from these documents is a complex, but solvable task.

Unstructured documents (letters, contracts, articles, etc.) could be flexible with structure and appearance.

The Internet and The Future

The idea is as simple as its application is difficult. If all cans, books, shoes or parts of cars are equipped with minuscule identifying devices, daily life on our planet will undergo a transformation. Things like running out of stock or wasted products will no longer exist as we will know exactly what is being consumed on the other side of the globe. Theft will be a thing of the past as we will know where a product is at all times. Counterfeiting of critical or costly items such as drugs, repair parts, or electronic components will be reduced or eliminated because manufacturers or other supply chain entities will know where their products are at all times. Product wastage or spoilage will be reduced because environmental sensors will alert suppliers or consumers when sensitive products are exposed to excessive heat, cold, vibration, or other risks. Supply chains will operate far more efficiently because suppliers will ship only the products needed when and where they are needed. Consumer and supplier prices should also drop accordingly.

The global association Auto-ID Center was founded in 1999 and is made up of 100 of the largest companies in the world such as Wal-Mart, Coca-Cola, Gillette, Johnson & Johnson, Pfizer, Procter & Gamble, Unilever, UPS, companies working in the sector of technology such as SAP, Aliens, Sun as well as five academic research centers. These are based at the following Universities; MIT in the USA, Cambridge University in the UK, the University of Adelaide in Australia, Keio University in Japan and University of St. Gallen in Switzerland.

The Auto-ID Center suggests a concept of a future supply chain that is based on the Internet of objects, i.e. a global application of RFID. They try to harmonize technology, processes and organization. Research is focused on miniaturization (aiming for a size of 0.3 mm/chip), reduction in the price per single device (aiming at around $0.05 per unit), the development of innovative application such as payment without any physical contact (Sony/Philips), domotics (clothes equipped with radio tags and intelligent washing machines), and sporting events (timing at the Berlin marathon).

AIDC 100

AIDC 100 is a professional organization for the automatic identification and data capture (AIDC) industry. This group is composed of individuals who made substantial contributions to the advancement of the industry. Increasing business's understanding of AIDC processes and technologies are the primary goals of the organization.

ISO/IEC 20248

ISO/IEC 20248 Automatic Identification and Data Capture Techniques – Data Structures – Digital Signature Meta Structure is an international standard specification under development by ISO/IEC JTC1 SC31 WG2. This development is an extension of SANS 1368, which is the current published specification. ISO/IEC 20248 and SANS 1368 are equivalent standard specifications. SANS 1368 is a South African national standard developed by the South African Bureau of Standards.

ISO/IEC 20248 [and SANS 1368] specifies a method whereby data stored within a barcode and/or RFID tag is structured and digitally signed. The purpose of the standard is to provide an open and interoperable method, between services and data carriers, to verify data originality and data integrity in an offline use case. The ISO/IEC 20248 data structure is also called a "DigSig" which refers to a small, in bit count, digital signature.

ISO/IEC 20248 also provides an effective and interoperable method to exchange data messages in the Internet of Things [IoT] and machine to machine [M2M] services allowing intelligent agents in such services to authenticate data messages and detect data tampering.

Description

ISO/IEC 20248 can be viewed as an X.509 application specification similar to S/MIME. Classic digital signatures are typically too big (the digital signature size is typically more than 2k bits) to fit in barcodes and RFID tags while maintaining the desired read performance. ISO/IEC 20248 digital signatures, including the data, are typically smaller than 512 bits. X.509 digital certificates within a public key infrastructure (PKI) is used for key and data description distribution. This method ensures the open verifiable decoding of data stored in a barcode and/or RFID tag into a tagged data structure; for example JSON and XML.

ISO/IEC 20248 addresses the need to verify the integrity of physical documents and objects. The standard counters verification costs of online services and device to server malware attacks by providing a method for multi-device and offline verification of the data structure. Examples documents and objects are education and medical certificates, tax and share/stock certificates, licences, permits, contracts, tickets, cheques, border documents, birth/death/identity documents, vehicle registration plates, art, wine, gemstones and medicine.

A DigSig stored in a QR code or near field communications (NFC) RFID tag can easily be read and verified using a smartphone with a ISO/IEC 20248 compliant application. The application only need to go online once to obtain the appropriate DigSig certificate, where after it can offline verify all DigSigs generated with that DigSig certificate.

A DigSig stored in a barcode can be copied without influencing the data verification. For example; a birth or school certificate containing a DigSig barcode can be copied. The copied document can also be verified to contain the correct information and the issuer of the information. A DigSig barcode provides a method to detect tampering with the data.

A DigSig stored in an RFID/NFC tag provides for the detection of copied and tampered data, therefore it can be used to detect the original document or object. The unique identifier of the RFID tag is used for this purpose.

The DigSig Envelope

ISO/IEC 20248 calls the digital signature meta structure a DigSig envelope. The DigSig envelope structure contains the DigSig certificate identifier, the digital signature and the timestamp. Fields can be contained in a DigSig envelope in 3 ways; Consider the envelope DigSig{a, ƀ, c} which contains field sets a, ƀ and c.

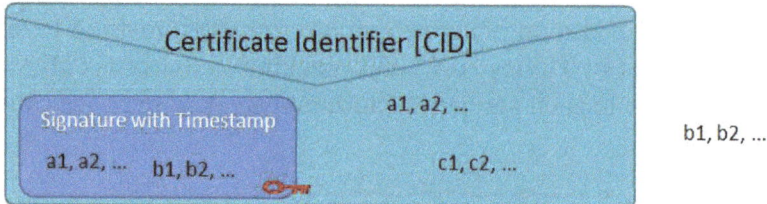

a fields are signed and included in the DigSig envelope. All the information (the signed field value and the field value is stored on the AIDC) is available to verify when the data structure is read from the AIDC (barcode and/or RFID).

ƀ fields are signed but NOT included in the DigSig envelope - only the signed field value is stored on the AIDC. Therefor the value of a ƀ field must be collected by the verifier before verification can be performed. This is useful to link a physical object with an barcode and/or RFID tag to be used as an anti-counterfeiting measure; for example the seal number of a bottle of wine may be a ƀ field. The verifier needs to enter the seal number for a successful verification since it is not stored in the barcode on the bottle. When the seal is broken the seal number may also be destroyed and yielded unreadable; the verification can therefore not take place since it requires the seal number. A replacement seal must display the same seal number; using holograms and other techniques may make the generation of a new copied seal number not viable. Similarly the unique tag ID, also known is the TID in ISO/IEC 18000, can be used in this manner to proof that the data is stored on the correct tag. In this case the TID is a ƀ field. The interrogator will read the DigSig envelope from the changeable tag memory and then read the non-changeable unique TID to allow for the verification. If the data was copied from one tag to another, then the verification process of the signed TID, as stored in the DigSig envelope, will reject the TID of the copied tag.

c fields are NOT signed but included in the DigSig envelope - only the field value is stored on the AIDC. A c field can therefore NOT be verified, but extracted from the AIDC. This field value may be changed without affecting the integrity of the signed fields.

The DigSig Data Path

Typically data stored in a DigSig originate as structured data; JSON or XML. The structured data field names maps directly on the DigSig Data Description [DDD].

This allows the DigSig Generator to digitally sign the data, store it in the DigSig envelope and compact the DigSig envelope to fit in the smallest bits size possible. The DigSig envelope is then programmed in an RFID tag or printed within a barcode symbology.

The DigSig Verifier reads the DigSig envelope from the barcode or RFID tag. It then identifies the relevant DigSig certificate, which it uses to extract the fields from the DigSig envelope and obtain the external fields. The Verifier then performs the verification and makes the fields available as structured data for example JSON or XML.

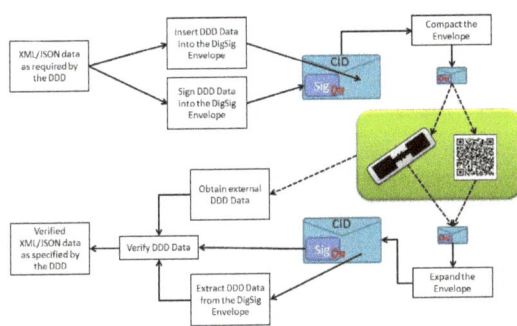

Examples

QR Example

The following education certificate examples use the URI-RAW DigSig envelope format. The URI format allows a generic barcode reader to read the DigSig where after it can be verified online using the URI of the trusted issuer of the DigSig. Often the ISO/IEC 20248 compliant smartphone application (App) will be available on this website for down load, where after the DigSig can be verified offline. Note, a compliant App must be able to verify DigSigs from any trusted DigSig issuer.

The university certificate example illustrates the multi-language support of SANS 1368.

CEM

**Centre for
Environmental Management**

This is to certify that

PT Pompies
1234567890123

successfully completed the course

**Introduction to Environmental Management -
An Overview of Principles, Tools and Issues**

CEM-01.1/0001/2011   15-19 September 2012   NQF Level: 5

Prof. JG Nel
Executive Manager:
Centre for Environmental Management
Course Leader

Prof. IJ Pienaar
Dean Faculty of Natural Science

University certificate example

RFID and QR Example

In this example a vehicle registration plate is fitted with an ISO/IEC 18000-63 (Type 6C) RFID tag and printed with a QR barcode. The plate is both offline verifiable using a smartphone, when the vehicle is stopped; or using an RFID reader, when the vehicle drive past the reader.

Note the 3 DigSig Envelope formats; RAW, URI-RAW and URI-TEXT.

The DigSig stored in the RFID tag is typically in a RAW envelope format to reduce the size from the URI envelope format. Barcodes will typically use the URI-RAW format to allow generic barcode readers to perform an online verification. The RAW format is the most compact but it can only be verified with a SANS 1368 compliant application.

This example uses both a QR and UHF RFID (6C). The DigSig stored on the 6C tag also contains the TID. Copying of the data can be detected and verified offline.

The same data is represented in a:
- DigSig RAW envelope
- DigSig URI-RAW envelope
- DigSig URI-TEXT envelope

DigSig RAW can only be read and verified with SANS 1358 compliant software.

http://sbox.idoctrust.com/verify/?C=2814&B=IEsA
R6YLM7s9zXa4mqymFVtROXJF0mWAtIssAooAssA=

http://sbox.idoctrust.com/verify?C=2814&
D=[["IEsAR6YLM7s9zXa4mqymFVtKOXJ=",
1349481600],[1234567890,"AA 99 AA"]]

The DigSig stored in the RFID tag will also contain the TID (Unique Tag Identifier) within the signature part. A DigSig Verifier will therefore be able to detect data copied onto another tag.

QR with External Data Example

The following QR barcode is attached to a computer or smartphone to prove it belongs to a specific person. It uses a ♭ type field, described above, to contain a secure personal identification number [PIN] remembered by the owner of the device. The DigSig Verifier will ask for the PIN to be entered, before the verification can take place. The verification will be negative if the PIN is incorrect. The PIN for the example is "123456".

SANS1368 QR - Equipment Permit with PIN

ISO/IEC JTC 1/SC 31 Automatic Identification and Data Capture Techniques

ISO/IEC JTC 1/SC 31 Automatic identification and data capture techniques is a standardization subcommittee of the joint subcommittee ISO/IEC JTC 1 of the International Organization for Standardization (ISO) and the International Electrotechnical Commission (IEC), that develops and facilitates international standards, technical reports, and technical specifications in the field of automatic identification and data capture techniques. The subcommittee had its first plenary in 1996, where it established its original three working groups. Additional working groups were established at subsequent plenaries such as ISO/IEC JTC 1/SC 31 WG 4 RFID, which was established at the third plenary in January 1998. The international national secretariat of ISO/IEC JTC 1/SC 31 is the American National Standards Institute (ANSI) located in the United States.

Scope

The scope of ISO/IEC JTC 1/SC 31 is "Standardization of data formats, data syntax, data structures, data encoding, and technologies for the process of automatic identification and data capture and of associated devices utilized in inter-industry applications and international business interchanges and for mobile applications."

Structure

ISO/IEC JTC 1/SC 31 is made up of, and six active working groups (WGs), each of which carries out specific tasks in standards development within the field of automatic identification and data capture techniques. Working groups can be disbanded if the group's working area is no longer applicable to standardization needs, or established if new working areas arise. The focus of each working group is described in the group's terms of reference. Active working groups of ISO/IEC JTC 1/ SC 31 are:

Working Group	Working Area	Status
ISO/IEC JTC 1/SC 31/ WG 1	Data carrier	Active
ISO/IEC JTC 1/SC 31/ WG 2	Data structure	Active
ISO/IEC JTC 1/SC 31/ WG 3	Conformance and performance	Disbanded - moved into WG 1 and WG 4
ISO/IEC JTC 1/SC 31/ WG 4	Radio communications (RFID, RTLS, Security)	Active
ISO/IEC JTC 1/SC 31/ WG 5	Real time location systems	Disbanded - moved into WG 4
ISO/IEC JTC 1/SC 31/ WG 6	Mobile item identification and management	Disbanded - moved into WG 2 and WG 4
ISO/IEC JTC 1/SC 31/ WG 7	Security and file management	Disbanded - moved into WG 4

Collaborations

ISO/IEC JTC 1/SC 31 works in close collaboration with a number of other organizations or subcommittees, both internal and external to ISO or IEC, in order to avoid conflicting or duplicative work. Organizations internal to ISO or IEC that collaborate with or are in liaison to ISO/IEC JTC 1/SC 31 include:

- ISO/IEC JTC 1/SC 6, Telecommunications and information exchange between systems * ISO/IEC JTC 1/SC 17, Cards and personal identification * ISO/IEC JTC 1/SC 17/WG 8, Integrated circuit cards without contacts * ISO/IEC JTC 1/SC 27, IT security techniques * ISO/IEC JTC 1/SC 37, Biometrics * ISO/IEC JTC 1/WG 7, Sensor networks * ISO/PC 246, Anti-counterfeiting tools * ISO/TC 104, Freight containers * ISO/TC 122, Packaging * ISO/TC 184/SC 4, Industrial data * ISO/TC 204, Intelligent transportation system * ISO/TC 247, Fraud countermeasures and controls

Some organizations external to ISO or IEC that collaborate with or are in liaison to ISO/IEC JTC 1/SC 31, include:

- AIM Global Inc., Association for automatic identification and mobility *CENELEC TC 106X, electromagnetic fields in the human environment *CEN/TC 225, AIDC technologies *CEN/TC 310, Advanced manufacturing technologies *CEN/TC 331, Postal services *Ecma International *European

Telecommunications Standards Institute (ETSI) *International Air Transport Association (IATA) *Institute of Electrical and Electronics Engineers, (IEEE) *ITU *GS1, GS1 Global Office/EPC Global *Universal Postal Union (UPU)

Member Countries

Countries pay a fee to ISO to be members of subcommittees. The 31 "P" (participating) members of ISO/IEC JTC 1/SC 31 are: Australia, Austria, Belgium, Brazil, Canada, China, Colombia, Czech Republic, Denmark, France, Germany, India, Ireland, Israel, Japan, Kenya, Republic of Korea, Malaysia, Netherlands, Peru, Philippines, Russian Federation, Singapore, Slovakia, South Africa, Spain, Sweden, Switzerland, United Kingdom, and United States. The 12 "O" (observer) members of ISO/IEC JTC 1/SC 31 are: Bosnia and Herzegovina, Finland, Ghana, Hong Kong, Hungary, Indonesia, Islamic Republic of Iran, Italy, Kazakhstan, Luxembourg, New Zealand, Romania, Serbia, and Thailand.

Standards

ISO/IEC JTC 1/SC 31 currently has 107 published standards within the field of automatic identification and data capture, including:

ISO/IEC Standard	Title	Status	Description	WG
ISO/IEC 15420	Information technology – Automatic identification and data capture techniques – EAN/UPC bar code symbology specification	Published (2009)	Specifies the requirements for the bar code symbology known as EAN/UPC; for use by manufacturers of bar code equipment and users of bar code technology	1
ISO/IEC 18004	Information technology – Automatic identification and data capture techniques – QR Code 2005 bar code symbology specification	Published (2006)	Defines the requirements for the QR Code 2005 symbology, by specifying its characteristics, data character encoding methods, symbol formats, dimensional characteristics, error correction rules, reference decoding algorithm, production quality requirements, and user-selectable application parameters, and by listing, in an informative annex, the features of QR Code Model 1 symbols, which differ from QR Code 2005	1

ISO/IEC 24728	Information technology – Automatic identification and data capture techniques – MicroPDF417 bar code symbology specification	Published (2006)	Specifies the requirements for the bar code symbology known as Micro-PDF417, and specifies its symbology characteristics, data character encodation, symbol formats, dimensions, error correction rules, decoding algorithm, and a number of application parameters	1
ISO/IEC 24778	Information technology – Automatic identification and data capture techniques – Aztec Code bar code symbology specification	Published (2008)	Defines the requirements for the symbology of Aztec Code, a two-dimensional matrix symbology whose symbols are nominally squared, made up of square modules on a square grid, with a square bulls-eye pattern at its center	1
ISO/IEC 15434	Example	Published (2006)	"Defines the manner in which data is transferred to high-capacity automatic data capture (ADC) media from a supplier's information system and the manner in which data is transferred to the recipient's information system."	2
ISO/IEC 15459	Information technology – Automatic identification and data capture techniques – Unique identification – Part 1 Individual transport units	Published (2006)	"Specifies a unique, non-significant string of characters for the identification of transport units. The character string is intended to be represented in a bar code label or other AIDC media attached to the item to meet item management needs. To address management needs, different classes of items are recognized in the various parts of ISO/IEC 15459, which allows different requirements to be met by the unique identifiers associated with each class. The rules for the unique identifier for transport units, to identify physical logistical transfers, with the identity relevant for the duration of one or more items in the load being held or transported as part of that load, are defined and supported by an example."	2

ISO/IEC 15418	Information technology – Automatic identification and data capture techniques – GS1 Application Identifiers and ASC MH10 Data Identifiers and maintenance	Published (2009)	Specifies sets of Data Identifiers and Application Identifiers for the purpose of identifying encoded data, and identifies the organizations responsible for their maintenance.	2
ISO/IEC 20248	Automatic Identification and Data Capture Techniques – Data Structures – Digital Signature Meta Structure	Published as SANS 1368, ISO/IEC draft under development	Specifies a method whereby data stored within a barcode and/or RFID tag is structured and digitally signed. The purpose of the standard is to provide an open and interoperable method, between services and data carriers, to verify data originality and data integrity in an offline use case.	2
ISO/IEC 15961-1	Information technology – Radio-frequency identification (RFID) for item management: Data protocol – Part 1: Application interface	Published (2013)	<ul><li>Provides guidelines on how data shall be presented as objects</li><li>Defines the structure of Object Identifiers</li><li>Specifies the commands that are supported for transferring data between an application and the RFID tag</li><li>Specifies the responses that are supported for transferring data between the RFID tag and the application</li></ul>	4
ISO/IEC 15963	Information technology – Radio frequency identification for item management – Unique identification for RF tags	Published (2009)	Describes numbering systems that are available for the identification RF tags	4

ISO/IEC 18000	Information technology – Radio frequency identification for item management –	Published (2008)	The various parts of ISO/IEC 18000 describe air interface communication at different frequencies in order to be able to utilize the different physical behaviors. The various parts of ISO/IEC 18000 are developed by ISO/IEC JTC1 SC31, "Automatic Data Capture Techniques".	4
ISO/IEC 18046	Information technology – Radio frequency identification device performance test methods – Part 1: Test methods for system performance	Published (2011)	Defines test methods for the performance characteristics of RFID systems for item management; specifies the general requirements and test requirements for systems which are applicable to the selection of devices for an application	4
ISO/IEC 24730-61	Information technology— Real time locating systems (RTLS) -- Part 61: Low rate pulse repetition frequency Ultra Wide Band (UWB) air interface	Published (2013)	Defines the physical layer (PHY) and tag management layer (TML) of an ultra wide band (UWB) air interface protocol that supports one directional simplex communication readers and tags of a real time locating system (RTLS). This protocol is best utilized for low-data-rate wireless connectivity with fixed, portable, and moving devices with very limited battery consumption requirements.	
ISO/IEC 24730-62	Information technology— Real time locating systems (RTLS) -- Part 62: High rate pulse repetition frequency Ultra Wide Band (UWB) air interface	Published (2013)	Defines the air-interface for real time locating systems (RTLS) using a physical layer Ultra Wide Band (UWB) signalling mechanism (based on IEEE 802.15.4a UWB). This modulation scheme employs high rate pulse repetition frequencies (PRF) 16 MHz or 64 MHz, and a combination of burst position modulation (BPM) and binary phase-shift keying (BPSK) giving an extremely high level of performance with a fully coherent receiver.	

ISO/IEC 29143	Information technology— Real time locating systems (RTLS) -- Part 62: High rate pulse repetition frequency Ultra Wide Band (UWB) air interface	Published (2013)	Defines the air-interface for real time locating systems (RTLS) using a physical layer Ultra Wide Band (UWB) signalling mechanism (based on IEEE 802.15.4a UWB). This modulation scheme employs high rate pulse repetition frequencies (PRF) 16 MHz or 64 MHz, and a combination of burst position modulation (BPM) and binary phase-shift keying (BPSK) giving an extremely high level of performance with a fully coherent receiver.	
ISO/IEC 29143	Information technology – Automatic identification and data capture techniques – Air interface specification for Mobile RFID interrogators	Published (2011)	Provides an air interface specification for Mobile RFID interrogators being part of a passive backscatter system	6
ISO/IEC/ IEEE 21451-1	Information technology – Smart transducer interface for sensors and actuators – Part 1: Network Capable Application Processor (NCAP) information model	Published (2010)	"Defines an object model with a network-neutral interface for connecting processers to communication networks, sensors, and actuators."	6
ISO/IEC 29167-1	Information technology – Automatic identification and data capture techniques – Part 1: Air interface for security services and file	Published (2012)	Defines the architecture for security and file management for the ISO/IEC 18000 air interface standards for RFID devices, and extends the air interface through the definition of architecture for: • Untraceability • Security services	7

Standards currently under development by ISO/IEC JTC 1/SC 31 include standards for Optical Character Recognition (OCR) by ISO/IEC JTC 1/SC W31/WG 1, standards for bar code symbols on mobile phone displays, and reading and display of ORM by mobile devices.

Major Working Areas of Data Capture Techniques

Mobile Phone Tracking

Mobile phone tracking is the ascertaining of the position or location of a mobile phone, whether stationary or moving. Localization may occur either via multilateration of radio signals between (several) cell towers of the network and the phone, or simply via GPS. To locate a mobile phone using multilateration of radio signals, it must emit at least the roaming signal to contact the next nearby antenna tower, but the process does not require an active call. The Global System for Mobile Communications (GSM) is based on the phone's signal strength to nearby antenna masts.

Mobile positioning may include location-based services that disclose the actual coordinates of a mobile phone, which is a technology used by telecommunication companies to approximate the location of a mobile phone, and thereby also its user.

Technology

The technology of locating is based on measuring power levels and antenna patterns and uses the concept that a powered mobile phone always communicates wirelessly with one of the closest base stations, so knowledge of the location of the base station implies the cell phone is nearby.

Advanced systems determine the sector in which the mobile phone is located and roughly estimate also the distance to the base station. Further approximation can be done by interpolating signals between adjacent antenna towers. Qualified services may achieve a precision of down to 50 meters in urban areas where mobile traffic and density of antenna towers (base stations) is sufficiently high. Rural and desolate areas may see miles between base stations and therefore determine locations less precisely.

GSM localization uses multilateration to determine the location of GSM mobile phones, or dedicated trackers, usually with the intent to locate the user.

The location of a mobile phone can be determined in a number of ways:

Network-based

The location of a mobile phone can be determined using the service provider's network infrastructure. The advantage of network-based techniques, from a service provider's point of view, is that they can be implemented non-intrusively without affecting handsets. Network-based techniques were developed many years prior to the widespread availability of GPS on handsets.

The accuracy of network-based techniques varies, with cell identification as the least accurate and triangulation as moderately accurate, and newer "advanced forward link trilateration" timing methods as the most accurate. The accuracy of network-based techniques is both dependent on the concentration of cell base stations, with urban environments achieving the highest possible accuracy because of the higher number of cell towers, and the implementation of the most current timing methods.

One of the key challenges of network-based techniques is the requirement to work closely with the service provider, as it entails the installation of hardware and software within the operator's infrastructure. Frequently the compulsion associated with a legislative framework, such as Enhanced 9-1-1, is required before a service provider will deploy a solution.

Handset-based

The location of a mobile phone can be determined using client software installed on the handset. This technique determines the location of the handset by putting its location by cell identification, signal strengths of the home and neighboring cells, which is continuously sent to the carrier. In addition, if the handset is also equipped with GPS then significantly more precise location information can be then sent from the handset to the carrier.

Another approach is to use a fingerprinting-based technique, where the "signature" of the home and neighboring cells signal strengths at different points in the area of interest is recorded by war-driving and matched in real-time to determine the handset location. This is usually performed independent from the carrier.

The key disadvantage of handset-based techniques, from service provider's point of view, is the necessity of installing software on the handset. It requires the active cooperation of the mobile subscriber as well as software that must be able to handle the different operating systems of the handsets. Typically, smartphones, such as one based on Symbian, Windows Mobile, Windows Phone, BlackBerry OS, iOS, or Android, would be able to run such software, e.g. Google Maps.

One proposed work-around is the installation of embedded hardware or software on the handset by the manufacturers, e.g., Enhanced Observed Time Difference (E-OTD). This avenue has not made significant headway, due to the difficulty of convincing different manufacturers to cooperate on a common mechanism and to address the cost issue. Another difficulty would be to address the issue of foreign handsets that are roaming in the network.

SIM-based

Using the subscriber identity module (SIM) in GSM and Universal Mobile Telecommunications System (UMTS) handsets, it is possible to obtain raw radio measurements from the handset. Available measurements include the serving Cell ID, round-trip time, and signal strength. The type of information obtained via the SIM can differ from that which is available from the handset. For example, it may not be possible to obtain any raw measurements from the handset directly, yet still obtain measurements via the SIM.

Wi-Fi

Crowdsourced Wi-Fi data can also be used to identify a handset's location. Poor performance of the GPS-based methods in indoor environment and increasing popularity of Wi-Fi have encouraged companies to design new and feasible methods to carry out Wi-Fi-based indoor positioning. Most smartphones combine Global Navigation Satellite Systems (GNSS), such as GPS and GLONASS, with Wi-Fi positioning systems.

Hybrid

Hybrid positioning systems use a combination of network-based and handset-based technologies for location determination. One example would be some modes of Assisted GPS, which can both use GPS and network information to compute the location. Both types of data are thus used by the telephone to make the location more accurate (i.e., A-GPS). Alternatively tracking with both systems can also occur by having the phone attain its GPS-location directly from the satellites, and then having the information sent via the network to the person that is trying to locate the telephone. Such systems include Google Maps, as well as, LTE's OTDOA and E-CellID.

There are also hybrid positioning systems which combine several different location approaches to position mobile devices by Wi-Fi, WiMAX, GSM, LTE, IP addresses, and network environment data.

Operational Purpose

In order to route calls to a phone, the cell towers listen for a signal sent from the phone and negotiate which tower is best able to communicate with the phone. As the phone changes location, the antenna towers monitor the signal, and the phone is "roamed" to an adjacent tower as appropriate. By comparing the relative signal strength from multiple antenna towers, a general location of a phone can be roughly determined. Other means make use of the antenna pattern, which supports angular determination and phase discrimination.

Newer phones may also allow the tracking of the phone even when turned on and not active in a telephone call. This results from the roaming procedures that perform handover of the phone from one base station to another.

Bearer Interest

A phone's location can be uploaded to a common website where one's friends and family can view one's last reported position. Newer phones may have built-in GPS receivers which could be used in a similar fashion, but with much higher accuracy. This is controversial, because data on a common website means people who are not "friends and family" may be able to view the information.

Privacy

Locating or positioning touches upon delicate privacy issues, since it enables someone to check where a person is without the person's consent. Strict ethics and security measures are strongly recommended for services that employ positioning. In 2012 Malte Spitz held a TED talk on the issue of mobile phone privacy in which he showcased his own stored data that he received from Deutsche Telekom after suing the company. He described the data, which consists of 35,830 lines of data collected during the span of Germany's data retention at the time, saying, "This is six months of my life [...] You can see where I am, when I sleep at night, what I'm doing." He partnered up with ZEIT Online and made his information publicly available in an interactive map which allows users to watch his entire movements during that time in fast-forward. Spitz concluded that technology consumers are the key to challenging privacy norms in today's society who "have to fight for self determination in the digital age."

China

China has proposed using this technology to track commuting patterns of Beijing city residents. Aggregate presence of mobile phone users could be tracked in a privacy-preserving fashion.

Europe

In Europe most countries have a constitutional guarantee on the secrecy of correspondence, and location data obtained from mobile phone networks is usually given the same protection as the communication itself.

United States

In the United States, there is no explicit constitutional guarantee on the privacy of telecommunications, so use of location data is limited by law. Law enforcement can obtain permission to position phones in emergencies where people, including criminals, are missing. In some instances, law enforcement may even access a mobile phone's internal microphone to eavesdrop on conversations while the phone is switched off.

A secret interpretation of The Patriot Act, confirmed to exist, has been linked to secret widespread location tracking.

Since 2005 the Electronic Frontier Foundation has been following some U.S. cases, including USA v. Pen Register, regarding government tracking of individuals. In In re Application of the United States for Historical Cell Site Data, 724 F.3d 600 (5th Cir. 2013), the United States Court of Appeals for the Fifth Circuit held that the government does not need a warrant to compel cell phone providers to disclose historical cell site information. However, in United States v. Davis (2014), the United States Court of Appeals for the Eleventh Circuit ruled in a criminal case that obtaining cell phone location data "without a warrant is a Fourth Amendment violation."

In 2014, it was revealed that in order to find fugitives, the United States Marshals Service has been flying small aircraft with equipment that identifies all cell phones in the area.

Commercial Privacy of Location Information in the United States

The U.S. does limit commercial use of location information under the (US) Telecommunications Act, at 47 CFR §222. The Telecommunications Act, at 47 CFR §222(f), requires consent from the subscriber, and prohibits telecommunication common carriers from accessing location information for purposes other than system operation without consent of the customer. Businesses such as Locaid, which provide a tracking service based on subscriber information, require mobile users' consent prior to tracking.

Indoor Positioning System

An indoor positioning system (IPS) is a system to locate objects or people inside a building using radio waves, magnetic fields, acoustic signals, or other sensory information collected by mobile devices. There are several commercial systems on the market, but there is no standard for an IPS system.

IPS systems use different technologies, including distance measurement to nearby anchor nodes (nodes with known positions, e.g., WiFi access points), magnetic positioning, dead reckoning. They either actively locate mobile devices and tags or provide ambient location or environmental context for devices to get sensed. The localized nature of an IPS has resulted in design fragmentation, with systems making use of various optical, radio, or even acoustic technologies.

System designs must take into account that at least three independent measurements are needed to unambiguously find a location. For smoothing to compensate for stochastic (unpredictable) errors there must be a sound method for reducing the error budget significantly. The system might include information from other systems to cope for physical ambiguity and to enable error compensation.

Applicability and Precision

Due to the signal attenuation caused by construction materials, the satellite based Global Positioning System (GPS) loses significant power indoors affecting the required coverage for receivers by at least four satellites. In addition, the multiple reflections at surfaces cause multi-path propagation serving for uncontrollable errors. These very same effects are degrading all known solutions for indoor locating which uses electromagnetic waves from indoor transmitters to indoor receivers. A bundle of physical and mathematical methods is applied to compensate for these problems. Promising direction radiofrequency positioning error correction opened by the use of alternative sources of navigational information, such as inertial measurement unit (IMU), monocular camera Simultaneous localization and mapping (SLAM) and WiFi SLAM. Integration of data from various navigation systems with different physical principles can increase the accuracy and robustness of the overall solution.

With detailed reading in the marketing documents and even in the specifications served by many of the IPS vendors, the interested customer will look for details on precision, reproducibility and other terms for quality of function with little success. Many vendors do not even tangle with the term accuracy

Relation to GPS

Global navigation satellite systems (GPS or GNSS) are generally not suitable to establish indoor locations, since microwaves will be attenuated and scattered by roofs, walls and other objects. However, in order to make positioning signals ubiquitous, integration between GPS and indoor positioning can be made.,

Currently, GNSS receivers are becoming more and more sensitive due to ceaseless progress in chip technology and processing power. High Sensitivity GNSS receivers are able to receive satellite signals in most indoor environments and attempts to determine the 3D position indoors have been successful. Besides increasing the sensitivity of the receivers, the technique of A-GPS is used, where the almanac and other information are transferred through a mobile phone.

However, proper coverage for the required four satellites to locate a receiver is not achieved with all current designs (2008–11) for indoor operations. Beyond, the average error budget for GNSS systems normally is much larger than the confinements, in which the locating shall be performed.

Locating and Positioning

Despite the name, most current IPS systems are so coarse that they cannot be used to detect the orientation or direction of an object.

Locating and Tracking

One of the methods to thrive for sufficient operational suitability is "tracking". Whether a sequence of locations determined form a trajectory from the first to the most actual location. Statistical methods then serve for smoothing the locations determined in a track resembling the physical capabilities of the object to move. This smoothing must be applied, when a target moves and also for a resident target, to compensate erratic measures. Otherwise the single resident location or even the followed trajectory would compose of an itinerant sequence of jumps.

Identification and Segregation

In most applications the population of targets is larger than just one. Hence the IPS must serve a proper specific identification for each observed target and must be capable to segregate and separate the targets individually within the group. An IPS must be able to identify the entities being tracked, despite the "non-interesting" neighbors. Depending on the design, either a sensor network must know from which tag it has received information, or a locating device must be able to identify the targets directly.

Non-radio Technologies

Non-radio technologies can be used for positioning without using the existing wireless infrastructure. This can provide increased accuracy at the expense of costly equipment and installations.

Magnetic Positioning

Magnetic positioning can offer pedestrians with smartphones an indoor accuracy of 1–2 meters with 90% confidence level, without using the additional wireless infrastructure for positioning. Magnetic positioning is based on the iron inside buildings that create local variations in the Earth's magnetic field. Un-optimized compass chips inside smartphones can sense and record these magnetic variations to map indoor locations.

Inertial Measurements

Pedestrian dead reckoning and other approaches for positioning of pedestrians propose an inertial measurement unit carried by the pedestrian either by measuring steps indirectly (step counting) or in a foot mounted approach, sometimes referring to maps or other additional sensors to constrain the inherent sensor drift encountered with inertial navigation. However, in order to make it capable to build map itself, the SLAM algorithm framework will be used.

Inertial measures generally cover the differentials of motion, hence the location gets determined with integrating and thus requires integration constants to provide results. The actual position estimation can be found as the maximum of a 2-d probability distribution which is recomputed at each step taking into account the noise model of all the sensors involved and the constraints posed by walls and furniture.

Wireless Technologies

Any wireless technology can be used for locating. Many different systems take advantage of existing wireless infrastructure for indoor positioning. There are three primary system topology options for hardware and software configuration, network-based, terminal-based, and terminal-assisted. Positioning accuracy can be increased at the expense of wireless infrastructure equipment and installations.

Wi-Fi-based Positioning System (WPS)

Wi-Fi positioning system (WPS) is used where GPS is inadequate. The localization technique used for positioning with wireless access points is based on measuring the intensity of the received signal (received signal strength in English RSS) and the method of "fingerprinting". Typical parameters useful to geolocate the WiFi hotspot or wireless access point include the SSID and the MAC address of the access point. The accuracy depends on the number of positions that have been entered into the database. The possible signal fluctuations that may occur can increase errors and inaccuracies in the path of the user. Anyplace is a free and open-source Wi-Fi positioning system that allows anybody to rapidly map indoor spaces and that won several awards for its location accuracy.

Bluetooth

According to the Bluetooth Special Interest Group, Bluetooth is all about proximity, not about exact location. Bluetooth was not intended to offer a pinned location like GPS, however is known as a geo-fence or micro-fence solution which makes it an indoor proximity solution, not an indoor positioning solution. Micromapping and indoor mapping has been linked to Bluetooth and to the Bluetooth LE based iBeacon promoted by Apple Inc.. Large-scale indoor positioning system based on iBeacons has been implemented and applied in practice.

Choke Point Concepts

Simple concept of location indexing and presence reporting for tagged objects, uses known sensor identification only. This is usually the case with passive radio-frequency identification (RFID) systems, which do not report the signal strengths and various distances of single tags or of a bulk of tags and do not renew any before known location coordinates of the sensor or current location of any tags. Operability of such approaches requires some narrow passage to prevent from passing by out of range.

Grid Concepts

Instead of long range measurement, a dense network of low-range receivers may be arranged, e.g. in a grid pattern for economy, throughout the space being observed. Due to the low range, a tagged entity will be identified by only a few close, networked receivers. An identified tag must be within range of the identifying reader, allowing a rough approximation of the tag location. Advanced systems combine visual coverage with a camera grid with the wireless coverage for the rough location.

Long range Sensor Concepts

Most systems use a continuous physical measurement (such as angle and distance or distance only) along with the identification data in one combined signal. Reach by these sensors mostly covers an entire floor, or an aisle or just a single room. Short reach solutions get applied with multiple sensors and overlapping reach.

Angle of Arrival

Angle of arrival (AoA) is the angle from which a signal arrives at a receiver. AoA is usually determined by measuring the time difference of arrival (TDOA) between multiple antennas in a sensor array. In other receivers, it is determined by an array of highly directional sensors—the angle can be determined by which sensor received the signal. AoA is usually used with triangulation and a known base line to find the location relative to two anchor transmitters.

Time of Arrival

Time of arrival (ToA, also time of flight) is the amount of time a signal takes to propagate from transmitter to receiver. Because the signal propagation rate is constant and known (ignoring differences in mediums) the travel time of a signal can be used to directly calculate distance. Multiple measurements can be combined with trilateration and multilateration to find a location. This is the technique used by GPS. Systems which use ToA, generally require a complicated synchronization mechanism to maintain a reliable source of time for sensors (though this can be avoided in carefully designed systems by using repeaters to establish coupling).

The accuracy of the TOA based methods often suffers from massive multipath conditions in indoor localization, which is caused by the reflection and diffraction of the RF signal from objects (e.g., interior wall, doors or furniture) in the environment. However, it is possible to reduce the effect of multipath by applying temporal or spatial sparsity based techniques.

Received Signal Strength Indication

Received signal strength indication (RSSI) is a measurement of the power level received by sensor. Because radio waves propagate according to the inverse-square law, distance can be approximated based on the relationship between transmitted and received signal strength (the transmission strength is a constant based on the equipment being used), as long as no other errors contribute to faulty results. The inside of buildings is not free space, so accuracy is significantly impacted by reflection and absorption from walls. Non-stationary objects such as doors, furniture, and people can pose an even greater problem, as they can affect the signal strength in dynamic, unpredictable ways.

A lot of systems use enhanced Wi-Fi infrastructure to provide location information. None of these systems serves for proper operation with any infrastructure as is. Unfortunately, Wi-Fi signal strength measurements are extremely noisy, so there is ongoing research focused on making more accurate systems by using statistics to filter out the inaccurate input data. Wi-Fi Positioning Systems are sometimes used outdoors as a supplement to GPS on mobile devices, where only few erratic reflections disturb the results.

Others

- Radio frequency identification (RFID): passive tags are very cost-effective, but do not support any metrics

- Ultrawide band (UWB): reduced interference with other devices

- Infrared (IR): previously included in most mobile devices

- Gen2IR (second generation infrared)

- Visible light communication (VLC): can use existing lighting systems

- Ultrasound: waves move very slowly, which results in much higher accuracy

Mathematics

Once sensor data has been collected, an IPS tries to determine the location from which the received transmission was most likely collected. The data from a single sensor is generally ambiguous and must be resolved by a series of statistical procedures to combine several sensor input streams.

Empirical Method

One way to determine position is to match the data from the unknown location with a large set of known locations using an algorithm such as k-nearest neighbor. This technique requires a comprehensive on-site survey and will be inaccurate with any significant change in the environment (due to moving persons or moved objects).

Mathematical Modeling

Location will be calculated mathematically by approximating signal propagation and finding angles and / or distance. Inverse trigonometry will then be used to determine location:

- Trilateration (distance from anchors)

- Triangulation (angle to anchors)

Advanced systems combine more accurate physical models with statistical procedures:

- Bayesian statistical analysis (probabilistic model)

- Kalman filtering (for estimating proper value streams under noise conditions).

Uses

The major consumer benefit of indoor positioning is the expansion of location-aware mobile computing indoors. As mobile devices become ubiquitous, contextual awareness for applications has become a priority for developers. Most applications currently rely on GPS, however, and function poorly indoors. Applications benefiting from indoor location include:

- Augmented reality

- School campus

- Museum guided tours.

- Shopping malls, including hypermarkets.

- Store navigation

- Warehouses

- Airports, bus, train and subway stations

- Parking lots, including these in hypermarkets

- Targeted advertising

- Social networking.

- Hospitals

- Hotels.

- Sports.

- Indoor robotics

- Tourism.

- Other public building maps.

Automatic Vehicle Location

Automatic vehicle location (AVL or ~locating; telelocating in EU) is a means for automatically determining and transmitting the geographic location of a vehicle. This data, from one or more vehicles, may then be collected by a vehicle tracking system for a picture of vehicle travel.

Most commonly, the location is determined using GPS, and the transmission mechanism is SMS, GPRS, a satellite or terrestrial radio from the vehicle to a radio receiver. GSM and EVDO are the most common services applied, because of the low data rate needed for AVL, and the low cost and near-ubiquitous nature of these public networks. The low bandwidth requirements also allow for satellite technology to receive telemetry data at a moderately higher cost, but across a global coverage area and into very remote locations not covered well by terrestrial radio or public carriers. Other options for determining actual location, for example in environments where GPS illumination is poor, are dead reckoning, i.e. inertial navigation, or active RFID systems or cooperative RTLS systems. With advantage, combinations of these systems may be applied. In addition, terrestrial radio positioning systems utilizing an LF (Low Frequency) switched packet radio network were also used as an alternative to GPS based systems.

Applications

Application with Vehicles

Automatic vehicle locating is a powerful concept for managing fleets of vehicles, as service vehicles, emergency vehicles, and especially precious construction equipment, also public transport vehicles (buses and trains). It is also used to track mobile assets, such as non wheeled construction equipment, non motorized trailers, and mobile power generators.

Application with Vehicle Drivers and Crews

The other purpose of tracking is to provide graded service or to manage a large driver and crewing staff effectively. For example, suppose an ambulance fleet has an objective of arriving at the location of a call for service within six minutes of receiving the request. Using an AVL system allows to evaluate the locations of all vehicles in service with driver and other crew in order to pick the vehicle that will most likely arrive at the destination fastest, (meeting the service objective).

Types of Systems

Simple Direction Finding

Amateur radio and some cellular or PCS wireless systems use direction finding or triangulation of transmitter signals radiated by the mobile. This is sometimes called radio direction finding or RDF. The simplest forms of these systems calculate the bearing from two fixed sites to the mobile. This creates a triangle with endpoints at the two fixed points and the mobile. Trigonometry tells you roughly where the mobile transmitter is located. In wireless telephone systems, the phones transmit continually when off-hook, making continual tracking and the collection of many location samples possible. This is one type of location system required by Federal Communications Commission Rules for wireless Enhanced 911.

Former LORAN-based Locating

Motorola offered a 1970s-era system based on the United States Coast Guard LORAN maritime navigation system. The LORAN system was intended for ships but signal levels on the US east- and west-coast areas were adequate for use with receivers in automobiles. The system may have been marketed under the Motorola model name Metricom. It consisted of an LF LORAN receiver and data interface box/modem connected to a separate two-way radio. The receiver and interface calculated a latitude and longitude in degrees, decimal degrees format based on the LORAN signals. This was sent over the radio as MDC-1200 or MDC-4800 data to a system controller, which plotted the mobile's approximate location on a map. The system worked reliably but sometimes had problems with electrical noise in urban areas. Sparking electric trolley poles or industrial plants which radiated electrical noise sometime overwhelmed the LORAN signals, affecting the system's ability to determine the mobile's geolocation. Because of the limited resolution, this type of system was impractical for small communities or operational areas such as a pit mine or port.

Signpost Systems

To track and locate vehicles along fixed routes, a technology called Signpost transmitters is employed. This is used on transit routes and rail lines where the vehicles to be tracked continually operated on the same linear route. A transponder or RFID chip along the vehicle route would be polled as the train or bus traverses its route. As each transponder was passed, the moving vehicle would query and receive an ack, or handshake, from the signpost transmitter. A transmitter on the mobile would report passing the signpost to a system controller. This allows supervision, a call center, or a dispatch center to monitor the progress of the vehicle and assess whether or not the vehicle was on schedule. These systems are an alternative inside tunnels or other conveyances where GPS signals are blocked by terrain.

Today's GPS-based Locating

The low price and ubiquity of Global Positioning System or GPS equipment has lent itself to more accurate and reliable telelocation systems. GPS signals are impervious to most electrical noise sources and don't require the user to install an entire system. Usually only a receiver to collect signals from the satellite segment is installed in each vehicle and radio or GSM to communicate the collected location data with a dispatch point.

Large private telelocation or AVL systems send data from GPS receivers in vehicles to a dispatch center over their private, user-owned radio backbone. These systems are used for businesses like parcel delivery and ambulances. Smaller systems which don't justify building a separate radio system use cellular or PCS data services to communicate location data from vehicles to their dispatching center. Location data is periodically polled from each vehicle in a fleet by a central controller or computer. In the simplest systems, data from the

GPS receiver is displayed on a map allowing humans to determine the location of each vehicle. More complex systems feed the data into a computer assisted dispatch system which automates the process. For example, the computer assisted dispatch system may check the location of a call for service and then pick a list of the four closest ambulances. This narrows the dispatcher's choice from the entire fleet to an easier choice of four vehicles.

Some wireless carriers such as Nextel have decided GPS was the best way to provide the mandated location data for wireless Enhanced 9-1-1. Newer Nextel radios have embedded GPS receivers which are polled if 9-1-1 is dialed. The 9-1-1 center is provided with latitude and longitude from the radio's GPS receiver. In centers with computer assisted dispatch, the system may assign an address to the call based on these coordinates or may project an icon depicting the caller's location onto a map of the area.

Sensor-augmented AVL

The main purpose of using AVL is not only to locate the vehicles, but also to obtain information about engine data, fuel consumption, driver data and sensor data from i.e. doors, freezer room on trucks or air pressure. Such data can be obtained via the CAN-bus, via direct connections to AVL systems or via open bus systems such as UFDEX that both sends and receives data via SMS or GPRS in pure ASCII text format. Because most AVL consists of two parts, GPS and GSM modem with additional embedded AVL software contained in a microcontroller, most AVL systems are fixed for its purposes unless they connect to an open bus system for expansion possibilities.

With an open bus system the users can send invoices based on goods delivered with exact location, time and date data where if connected to scale, RFID or barcode readers, can make a fairly good automated system to avoid human errors.

In countries with high prices on gasoline external fuel sensors are used to prevent cases of fuel theft.

Logbook Functions

Another scenario for sensor functions is to connect the AVL to driver information, to collect data about driving time, stops, or even driver absence from the vehicle. If the driver/worker conditions is such as the hourly rates for driving and working outside is not the same, this can be monitored by sensors, by using iButton or other personal identification devices. Later by analyzing log-file it is possible to get reports on any kind of events, like stops, visited streets, speed limits violations, etc.

Differentiating Between Automatic Vehicle Location and Events Activated Tracking Systems

It might be helpful to draw a distinction between vehicle location systems which track automatically and event activated tracking systems which track when triggered by an event.

There is increasingly crossover between the different systems and those with experience of this sector will be able to draw on a number of examples which break the rule.

A.V.L (Automatic Vehicle Location) This type of vehicle tracking is normally used in the fleet or driver management sector. The unit is configured to automatically transmit its location at a set time interval, e.g. every 5 minutes. The unit is activated when the ignition is switched on/off.

E.A.T.S (Events Activated Tracking system) This type of system is primarily used in connection with vehicle or driver security solutions. If, for example a thief breaks into your car and attempts to steal it, the tracking system can be triggered by the immobiliser unit or motion sensor being activated. A monitoring bureau, will then be automatically notified that the unit has been activated and begin tracking the vehicle.

Some products on the market are a hybrid of both AVL and EATS technology. However industry practice has tended to lean towards a separation of these functions. It is worth taking note that vehicle tracking products tend to fall into one, not both of the technologies.

AVL technology is predominately used when applying vehicle tracking to fleet or driver management solutions. The use of Automatic Vehicle Location is given in the following scenario; A car breaks down by the side of the road and the occupant calls a vehicle recovery company. The vehicle recovery company has several vehicles operating in the area. Without needing to call each driver to check his location the dispatcher can pinpoint the nearest recovery vehicle and assign it to the new job. If you were to incorporate the other aspects of vehicle telematics into this scenario; the dispatcher, rather than phoning the recovery vehicle operative, could transmit the job details directly to the operative's mobile data device, who would then use the in-vehicle satellite navigation to aid his journey to the job.

EATS technology is predominately used when applying vehicle tracking to vehicle security solutions. An example of this distinction is given in the following scenario; A construction company owns some pieces of plant machinery that are regularly left unattended, at weekends, on building sites. Thieves break onto one site and a piece equipment, such as a digger, is loaded on the back of a flat bed truck and then driven away. Typically the ignition wouldn't need to be turned on and as such most of the AVL products available wouldn't typically be activated. Only products that included a unit that was activated by a motion sensor or GeoFence alarm event, would be activated.

Both AVL and EATS systems track, but often for different purposes.

Special Applications of Automatic Vehicle Locating

Vehicle location technologies can be used in the following scenarios:

- Fleet management: when managing a fleet of vehicles, knowing the real-time location of all drivers allows management to meet customer needs more efficiently. Vehicle location information can also be used to verify that legal requirements are being met: for example, that drivers are taking rest breaks and obeying speed limits.

- Passenger Information: Real-time Passenger information systems use predictions based on AVL input to show the expected arrival and departure times of Public Transport services.

- Asset tracking: companies needing to track valuable assets for insurance or other monitoring purposes can now plot the real-time asset location on a map and closely monitor movement and operating status. For example, haulage and logistics companies often operate trucks with detachable load carrying units. In this case, trailers can be tracked independently of the cabs used to drive them. Combining vehicle location with inventory management that can be used to reconcile which item is currently on which vehicle can be used to identify physical location down to the level of individual packages.

- Field worker management: companies with a field service or sales workforce can use information from vehicle tracking systems to plan field workers' time, schedule subsequent customer visits and be able to operate these departments efficiently.

- Covert surveillance: vehicle location devices attached covertly by law enforcement or espionage organizations can be used to track journeys made by individuals who are under surveillance

References

- Waldner, Jean-Baptiste (2008). Nanocomputers and Swarm Intelligence. London: ISTE John Wiley & Sons. pp. 205–214. ISBN 1-84704-002-0.

- Fitzgerald, Britney (25 July 2012). "Malte Spitz's TED Talk Takes On Mobile Phone Privacy Debate (VIDEO)". Huffington Post. Retrieved 26 January 2016.

- "Apple Is Launching A Vast Project To Map The Inside Of Every Large Building It Can". Business Insider. Retrieved 2014-06-12.

- Biss, Chuck; Sharkey, Frank (2012-10-15).SC 31 Chairman's Presentation to the November 2012 JTC 1 Plenary Meeting in Jeju. Retrieved 2013-11-07.

- Alex Abdo (16 March 2012). "Government Confirms That It Has Secret Interpretation of Patriot Act Spy Powers". Aclu.org. Retrieved 2012-11-17.

- Jaffer, Jameel (2012-03-15). "Sens. Wyden and Udall Weigh in on ACLU Patriot Act FOIA Case". Aclu.org. Retrieved 2012-11-17.

- Savage, Charlie (2012-03-16). "Democratic Senators Warn About Use of Patriot Act". NYTimes.com. Retrieved 2012-11-17.

- Ackerman, Spencer (May 2011). "There's a Secret Patriot Act, Senator Says | Danger Room". Wired.com. Retrieved 2012-11-17.

- "Atlas Bugged: Why the "Secret Law" of the Patriot Act Is Probably About Location Tracking | Cato @ Liberty". Cato-at-liberty.org. 2011-05-27. Retrieved 2012-11-17.

- "Wyden Continues To Press Intelligence Officials About Tracking Americans Under 'Secret' Interpretation Of The Patriot Act". Techdirt. 2011-07-27. Retrieved 2012-11-17.

- "AIDC 100". AIDC 100: Professionals Who Excel in Serving the AIDC Industry. Archived from the original on 24 July 2011. Retrieved 2 August 2011.

- "Tracking a suspect by any mobile phone: Tracking SIM and handset". BBC News. 2005-08-03. Retrieved 2010-01-02.

Permissions

All chapters in this book are published with permission under the Creative Commons Attribution Share Alike License or equivalent. Every chapter published in this book has been scrutinized by our experts. Their significance has been extensively debated. The topics covered herein carry significant information for a comprehensive understanding. They may even be implemented as practical applications or may be referred to as a beginning point for further studies.

We would like to thank the editorial team for lending their expertise to make the book truly unique. They have played a crucial role in the development of this book. Without their invaluable contributions this book wouldn't have been possible. They have made vital efforts to compile up to date information on the varied aspects of this subject to make this book a valuable addition to the collection of many professionals and students.

This book was conceptualized with the vision of imparting up-to-date and integrated information in this field. To ensure the same, a matchless editorial board was set up. Every individual on the board went through rigorous rounds of assessment to prove their worth. After which they invested a large part of their time researching and compiling the most relevant data for our readers.

The editorial board has been involved in producing this book since its inception. They have spent rigorous hours researching and exploring the diverse topics which have resulted in the successful publishing of this book. They have passed on their knowledge of decades through this book. To expedite this challenging task, the publisher supported the team at every step. A small team of assistant editors was also appointed to further simplify the editing procedure and attain best results for the readers.

Apart from the editorial board, the designing team has also invested a significant amount of their time in understanding the subject and creating the most relevant covers. They scrutinized every image to scout for the most suitable representation of the subject and create an appropriate cover for the book.

The publishing team has been an ardent support to the editorial, designing and production team. Their endless efforts to recruit the best for this project, has resulted in the accomplishment of this book. They are a veteran in the field of academics and their pool of knowledge is as vast as their experience in printing. Their expertise and guidance has proved useful at every step. Their uncompromising quality standards have made this book an exceptional effort. Their encouragement from time to time has been an inspiration for everyone.

The publisher and the editorial board hope that this book will prove to be a valuable piece of knowledge for students, practitioners and scholars across the globe.

Index

www.ingramcontent.com/pod-product-compliance
Lightning Source LLC
Chambersburg PA
CBHW061933190326
41458CB00009B/2728